M000302343

Learning Apache Cassandra

Build an efficient, scalable, fault-tolerant, and highly-available data layer into your application using Cassandra

Mat Brown

BIRMINGHAM - MUMBAI

Learning Apache Cassandra

Copyright © 2015 Packt Publishing

All rights reserved. No part of this book may be reproduced, stored in a retrieval system, or transmitted in any form or by any means, without the prior written permission of the publisher, except in the case of brief quotations embedded in critical articles or reviews.

Every effort has been made in the preparation of this book to ensure the accuracy of the information presented. However, the information contained in this book is sold without warranty, either express or implied. Neither the author, nor Packt Publishing, and its dealers and distributors will be held liable for any damages caused or alleged to be caused directly or indirectly by this book.

Packt Publishing has endeavored to provide trademark information about all of the companies and products mentioned in this book by the appropriate use of capitals. However, Packt Publishing cannot guarantee the accuracy of this information.

First published: February 2015

Production reference: 1190215

Published by Packt Publishing Ltd.
Livery Place
35 Livery Street
Birmingham B3 2PB, UK.

ISBN 978-1-78398-920-1

www.packtpub.com

Credits

Author
Mat Brown

Reviewers
N. Bharadwaj

Mark Kerzner

Michael Klishin

Sanjay Sharma

Commissioning Editor
Akram Hussain

Acquisition Editor
Rebecca Youé

Content Development Editor
Ritika Singh

Sharvari Tawde

Technical Editor
Ruchi Desai

Copy Editors
Pranjali Chury

Merilyn Pereira

Project Coordinator
Judie Jose

Proofreaders
Simran Bhogal

Stephen Copestake

Kelly Hutchinson

Indexer
Monica Ajmera Mehta

Graphics
Sheetal Aute

Disha Haria

Production Coordinator
Conidon Miranda

Cover Work
Conidon Miranda

About the Author

Mat Brown is a professional software engineer in Brooklyn, New York. In his career, he has focused on building consumer-facing web and mobile applications for several companies; he currently works at Genius.

Mat is an enthusiastic contributor to the Ruby open source ecosystem. He is the maintainer of Cequel, a Ruby object mapper for Cassandra, and was the original author of Sunspot, a library that seamlessly integrates Solr search into Rails applications.

I would like to thank my wife, Pamela, and my parents for their love and support. I'd also like to thank my friends, especially those who seemed impressed when I told them I was writing a book. My cat, Taco, though not good for much, did keep me company during some writing sessions and thus deserves a mention here as well.

About the Reviewers

N. Bharadwaj is a software developer at Glassbeam Inc. He has over 12 years of experience in enterprise product development and managing development teams. His experience covers software design and development in the areas of network-data and application-data analysis. He counts Scala and Java among his primary development languages, and PostgreSQL and Cassandra among the databases he is experienced with. Bharadwaj's interests include building high scalability systems, distributed computing, and machine learning. He received his bachelor of engineering degree in electronics and communication systems from Visvesvaraya Technological University, Karnataka, in 2002 and his MBA from the Indian Institute of Management in 2012.

Mark Kerzner holds degrees in law, math, and computer science. He has been designing software for many years and has been working on Hadoop-based systems since 2008. He is the president of SHMsoft, a provider of Hadoop applications for various verticals, a cofounder of Elephant Scale, a training and consulting company, as well as the coauthor of the open source book, *Hadoop Illuminated*. His book *HBase Designed Patterns*, written together with Sujee Maniyam, was recently published by Packt Publishing.

I would like to acknowledge the help of my colleagues, in particular, Sujee Maniyam, and last but not least, my multitalented family.

Michael Klishin is an experienced software engineer primarily working on data infrastructure. A long-time open source contributor, he maintains over 50 open source projects.

Sanjay Sharma has been building enterprise-grade solutions in the software industry for around 16 years and using Big Data and Cloud technologies over the past 5 or 6 years to solve complex business problems. He has extensive experience with cardinal technologies, including Cassandra, Hadoop, Hive, MongoDB, MPP DW, and Java/J2EE/SOA, which allowed him to pioneer the LinkedIn group Hadoop India.

I would like to thank my employer—Impetus and iLabs, especially its R&D department, which invests in cutting-edge technologies and allowed me to become a pioneer in mastering Cassandra and Hadoop-like technologies early on.

Most importantly, I want to acknowledge my wonderful family—my wife and son—who have always supported and encouraged me in all my endeavors in life.

www.PacktPub.com

Support files, eBooks, discount offers, and more

For support files and downloads related to your book, please visit www.PacktPub.com.

Did you know that Packt offers eBook versions of every book published, with PDF and ePub files available? You can upgrade to the eBook version at www.PacktPub.com and as a print book customer, you are entitled to a discount on the eBook copy. Get in touch with us at service@packtpub.com for more details.

At www.PacktPub.com, you can also read a collection of free technical articles, sign up for a range of free newsletters and receive exclusive discounts and offers on Packt books and eBooks.

https://www2.packtpub.com/books/subscription/packtlib

Do you need instant solutions to your IT questions? PacktLib is Packt's online digital book library. Here, you can search, access, and read Packt's entire library of books.

Why subscribe?

- Fully searchable across every book published by Packt
- Copy and paste, print, and bookmark content
- On demand and accessible via a web browser

Free access for Packt account holders

If you have an account with Packt at www.PacktPub.com, you can use this to access PacktLib today and view 9 entirely free books. Simply use your login credentials for immediate access.

Table of Contents

Preface

The crop of distributed databases that has come to the market in recent years appeals to application developers for several reasons. Their storage capacity is nearly limitless, bounded only by the number of machines you can afford to spin up. Masterless replication makes them resilient to adverse events, handling even a complete machine failure without any noticeable effect on the applications that rely on them. Log-structured storage engines allow these databases to handle high-volume write loads without blinking an eye.

But compared to traditional relational databases, not to mention newer document stores, distributed databases are typically feature-poor and inconvenient to work with. Read and write functionality is frequently confined to simple key-value operations, with more complex operations demanding arcane map-reduce implementations. Happily, Cassandra provides all of the benefits of a fully-distributed data store while also exposing a familiar, user-friendly data model and query interface.

I first encountered Cassandra working on an application that stored our users' extended social graphs across a variety of services. With a hundred or so alpha users in the system, it became clear that, even at relatively modest traction, our storage needs would go beyond what our PostgreSQL database could comfortably handle. After surveying the landscape of horizontally scalable data stores, we decided to migrate to Cassandra because its table-based data structure seemed to provide an easy migration path from the application we had already built.

Our first deployment of Cassandra ported our previous PostgreSQL schema table-for-table. Only after taking the application to production did we come to realize that our expertise designing schemas for a relational world didn't map directly to a distributed store such as Cassandra. While we were happy to be storing tons of data at very high write volumes, it was clear we weren't getting maximum performance out of the database.

The story has a happy ending: after a rough initial launch, we went back to the drawing board and reworked our data model from the ground up to take advantage of Cassandra's strengths. With that new version deployed, Cassandra effortlessly handled our application scaling to hundreds of thousands of users' social graphs.

The goal of this book is to teach you the easy way what we learned the hard way: how to use Cassandra effectively, powerfully, and efficiently. We'll explore Cassandra's ins and outs by designing the persistence layer for a messaging service that allows users to post status updates that are visible to their friends. By the end of the book, you'll be fully prepared to build your own highly scalable, highly available applications.

What this book covers

Chapter 1, Getting Up and Running with Cassandra, introduces the major reasons to choose Cassandra over a traditional relational or document database. It then provides step-by-step instructions on installing Cassandra, creating a keyspace, and interacting with the database using the CQL language and cqlsh tool.

Chapter 2, The First Table, is a walk-through of creating a table, inserting data, and retrieving rows by primary key. Along the way, it discusses how Cassandra tables are structured, and provides a tour of the Cassandra type system.

Chapter 3, Organizing Related Data, introduces more complex table structures that group related data together using compound primary keys.

Chapter 4, Beyond Key-Value Lookup, puts the more robust schema developed in the previous chapter to use, explaining how to query for sorted ranges of rows.

Chapter 5, Establishing Relationships, develops table structures for modeling relationships between rows. The chapter introduces static columns and row deletion.

Chapter 6, Denormalizing Data for Maximum Performance, explains when and why storing multiple copies of the same data can make your application more efficient.

Chapter 7, Expanding Your Data Model, demonstrates the use of lightweight transactions to ensure data integrity. It also introduces schema alteration, row updates, and single-column deletion.

Chapter 8, Collections, Tuples, and User-defined Types, introduces collection columns and explores Cassandra's support for advanced, atomic collection manipulation. It also introduces tuples and user-defined types.

Chapter 9, Aggregating Time-Series Data, covers the common use case of collecting high-volume time-series data and introduces counter columns.

Chapter 10, How Cassandra Distributes Data, explores what happens when you save a row to Cassandra. It considers eventual consistency and teaches you how to use tunable consistency to get the right balance between consistency and fault-tolerance.

Appendix A, Peeking Under the Hood, peels away the abstractions provided by CQL to reveal how Cassandra represents data at the lower column family level.

Appendix B, Authentication and Authorization, introduces ways to control access to your Cassandra cluster and specific data structures within it.

What you need for this book

You will need the following software to work with the examples in this book:

- Java Runtime Environment 7.0 (`http://www.oracle.com/technetwork/java/javase/downloads/jre7-downloads-1880261.html`)
- Apache Cassandra 2.1 (`http://cassandra.apache.org/download/`)

Further instructions on installing these are present in the upcoming chapters.

Who this book is for

This book is for first-time users of Cassandra, as well as anyone who wants a better understanding of Cassandra in order to evaluate it as a solution for their application. Since Cassandra is a standalone database, we don't assume any particular coding language or framework; anyone who builds applications for a living, and who wants those applications to scale, will benefit from reading the book.

Conventions

In this book, you will find a number of text styles that distinguish between different kinds of information. Here are some examples of these styles and an explanation of their meaning.

Code words in text, database table names, folder names, filenames, file extensions, pathnames, dummy URLs, user input, and Twitter handles are shown as follows: "We can include other contexts through the use of the `include` directive."

A block of code is set as follows:

```
SELECT * FROM "users"
WHERE "username" = 'alice';
```

Any command-line input or output is written as follows:

```
$ sudo apt-get install cassandra
```

New terms and **important words** are shown in bold. Words that you see on the screen, for example, in menus or dialog boxes, appear in the text like this: "On this page, locate **Windows Server 2008/Windows 7 or Later (32-Bit)** from the Operating System menu (you might also want to look for 64-bit if you run a 64-bit version of Windows), and choose MSI Installer (2.x) from the version columns."

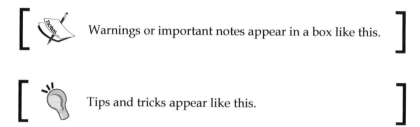

Warnings or important notes appear in a box like this.

Tips and tricks appear like this.

Reader feedback

Feedback from our readers is always welcome. Let us know what you think about this book—what you liked or disliked. Reader feedback is important for us as it helps us develop titles that you will really get the most out of.

To send us general feedback, simply e-mail feedback@packtpub.com, and mention the book's title in the subject of your message.

If there is a topic that you have expertise in and you are interested in either writing or contributing to a book, see our author guide at www.packtpub.com/authors.

Customer support

Now that you are the proud owner of a Packt book, we have a number of things to help you to get the most from your purchase.

Downloading the example code

You can download the example code files from your account at http://www.packtpub.com for all the Packt Publishing books you have purchased. If you purchased this book elsewhere, you can visit http://www.packtpub.com/support and register to have the files e-mailed directly to you.

Errata

Although we have taken every care to ensure the accuracy of our content, mistakes do happen. If you find a mistake in one of our books—maybe a mistake in the text or the code—we would be grateful if you could report this to us. By doing so, you can save other readers from frustration and help us improve subsequent versions of this book. If you find any errata, please report them by visiting `http://www.packtpub.com/submit-errata`, selecting your book, clicking on the **Errata Submission Form** link, and entering the details of your errata. Once your errata are verified, your submission will be accepted and the errata will be uploaded to our website or added to any list of existing errata under the Errata section of that title.

To view the previously submitted errata, go to `https://www.packtpub.com/books/content/support` and enter the name of the book in the search field. The required information will appear under the **Errata** section.

Piracy

Piracy of copyrighted material on the Internet is an ongoing problem across all media. At Packt, we take the protection of our copyright and licenses very seriously. If you come across any illegal copies of our works in any form on the Internet, please provide us with the location address or website name immediately so that we can pursue a remedy.

Please contact us at `copyright@packtpub.com` with a link to the suspected pirated material.

We appreciate your help in protecting our authors and our ability to bring you valuable content.

Questions

If you have a problem with any aspect of this book, you can contact us at `questions@packtpub.com`, and we will do our best to address the problem.

1
Getting Up and Running with Cassandra

As an application developer, you have almost certainly worked with databases extensively. You must have built products using relational databases like MySQL and PostgreSQL, and perhaps experimented with a document store like MongoDB or a key-value database like Redis. While each of these tools has its strengths, you will now consider whether a distributed database like Cassandra might be the best choice for the task at hand.

In this chapter, we'll talk about the major reasons to choose Cassandra from among the many database options available to you. Having established that Cassandra is a great choice, we'll go through the nuts and bolts of getting a local Cassandra installation up and running. By the end of this chapter, you'll know:

- When and why Cassandra is a good choice for your application
- How to install Cassandra on your development machine
- How to interact with Cassandra using cqlsh
- How to create a keyspace

What Cassandra offers, and what it doesn't

Cassandra is a fully distributed, masterless database, offering superior scalability and fault tolerance to traditional single master databases. Compared with other popular distributed databases like Riak, HBase, and Voldemort, Cassandra offers a uniquely robust and expressive interface for modeling and querying data. What follows is an overview of several desirable database capabilities, with accompanying discussions of what Cassandra has to offer in each category.

Horizontal scalability

Horizontal scalability refers to the ability to expand the storage and processing capacity of a database by adding more servers to a database cluster. A traditional single-master database's storage capacity is limited by the capacity of the server that hosts the master instance. If the data set outgrows this capacity, and a more powerful server isn't available, the data set must be **sharded** among multiple independent database instances that know nothing of each other. Your application bears responsibility for knowing to which instance a given piece of data belongs.

Cassandra, on the other hand, is deployed as a **cluster** of instances that are all aware of each other. From the client application's standpoint, the cluster is a single entity; the application need not know, nor care, which machine a piece of data belongs to. Instead, data can be read or written to any instance in the cluster, referred to as a **node**; this node will forward the request to the instance where the data actually belongs.

The result is that Cassandra deployments have an almost limitless capacity to store and process data; when additional capacity is required, more machines can simply be added to the cluster. When new machines join the cluster, Cassandra takes care of *rebalancing* the existing data so that each node in the expanded cluster has a roughly equal share.

Cassandra is one of the several popular distributed databases inspired by the Dynamo architecture, originally published in a paper by Amazon. Other widely used implementations of Dynamo include Riak and Voldemort. You can read the original paper at `http://s3.amazonaws.com/AllThingsDistributed/sosp/amazon-dynamo-sosp2007.pdf`.

High availability

The simplest database deployments are run as a single instance on a single server. This sort of configuration is highly vulnerable to interruption: if the server is affected by a hardware failure or network connection outage, the application's ability to read and write data is completely lost until the server is restored. If the failure is catastrophic, the data on that server might be lost completely.

A **master-follower** architecture improves this picture a bit. The master instance receives all write operations, and then these operations are **replicated** to follower instances. The application can read data from the master or any of the follower instances, so a single host becoming unavailable will not prevent the application from continuing to read data. A failure of the master, however, will still prevent the application from performing any write operations, so while this configuration provides high read availability, it doesn't completely provide high availability.

Cassandra, on the other hand, has no **single point of failure** for reading or writing data. Each piece of data is replicated to multiple nodes, but none of these nodes holds the authoritative master copy. If a machine becomes unavailable, Cassandra will continue writing data to the other nodes that share data with that machine, and will queue the operations and update the failed node when it rejoins the cluster. This means in a typical configuration, two nodes must fail simultaneously for there to be any application-visible interruption in Cassandra's availability.

How many copies?

When you create a keyspace—Cassandra's version of a database—you specify how many copies of each piece of data should be stored; this is called the **replication factor**. A replication factor of 3 is a common and good choice for many use cases.

Write optimization

Traditional relational and document databases are optimized for read performance. Writing data to a relational database will typically involve making in-place updates to complicated data structures on disk, in order to maintain a data structure that can be read efficiently and flexibly. Updating these data structures is a very expensive operation from a standpoint of disk I/O, which is often the limiting factor for database performance. Since writes are more expensive than reads, you'll typically avoid any unnecessary updates to a relational database, even at the expense of extra read operations.

Cassandra, on the other hand, is highly optimized for write throughput, and in fact never modifies data on disk; it only appends to existing files or creates new ones. This is much easier on disk I/O and means that Cassandra can provide astonishingly high write throughput. Since both writing data to Cassandra, and storing data in Cassandra, are inexpensive, **denormalization** carries little cost and is a good way to ensure that data can be efficiently read in various access scenarios.

Because Cassandra is optimized for write volume, you shouldn't shy away from writing data to the database. In fact, it's most efficient to *write without reading* whenever possible, even if doing so might result in redundant updates.

Just because Cassandra is optimized for writes doesn't make it bad at reads; in fact, a well-designed Cassandra database can handle very heavy read loads with no problem. We'll cover the topic of efficient data modeling in great depth in the next few chapters.

Structured records

The first three database features we looked at are commonly found in distributed data stores. However, databases like Riak and Voldemort are purely key-value stores; these databases have no knowledge of the internal structure of a record that's stored at a particular key. This means useful functions like updating only part of a record, reading only certain fields from a record, or retrieving records that contain a particular value in a given field are not possible.

Relational databases like PostgreSQL, document stores like MongoDB, and, to a limited extent, newer key-value stores like Redis do have a concept of the internal structure of their records, and most application developers are accustomed to taking advantage of the possibilities this allows. None of these databases, however, offer the advantages of a masterless distributed architecture.

In Cassandra, records are structured much in the same way as they are in a relational database—using tables, rows, and columns. Thus, applications using Cassandra can enjoy all the benefits of masterless distributed storage while also getting all the advanced data modeling and access features associated with structured records.

Secondary indexes

A secondary index, commonly referred to as an index in the context of a relational database, is a structure allowing efficient lookup of records by some attribute other than their primary key. This is a widely useful capability: for instance, when developing a blog application, you would want to be able to easily retrieve all of the posts written by a particular author. Cassandra supports secondary indexes; while Cassandra's version is not as versatile as indexes in a typical relational database, it's a powerful feature in the right circumstances.

Efficient result ordering

It's quite common to want to retrieve a record set ordered by a particular field; for instance, a photo sharing service will want to retrieve the most recent photographs in descending order of creation. Since sorting data on the fly is a fundamentally expensive operation, databases must keep information about record ordering persisted on disk in order to efficiently return results in order. In a relational database, this is one of the jobs of a secondary index.

In Cassandra, secondary indexes can't be used for result ordering, but tables can be structured such that rows are always kept sorted by a given column or columns, called **clustering columns**. Sorting by arbitrary columns at read time is not possible, but the capacity to efficiently order records in any way, and to retrieve ranges of records based on this ordering, is an unusually powerful capability for a distributed database.

Immediate consistency

When we write a piece of data to a database, it is our hope that that data is immediately available to any other process that may wish to read it. From another point of view, when we read some data from a database, we would like to be guaranteed that the data we retrieve is the most recently updated version. This guarantee is called **immediate consistency**, and it's a property of most common single-master databases like MySQL and PostgreSQL.

Distributed systems like Cassandra typically do not provide an immediate consistency guarantee. Instead, developers must be willing to accept **eventual consistency**, which means when data is updated, the system will reflect that update *at some point in the future*. Developers are willing to give up immediate consistency precisely because it is a direct tradeoff with high availability.

In the case of Cassandra, that tradeoff is made explicit through **tunable consistency**. Each time you design a write or read path for data, you have the option of immediate consistency with less resilient availability, or eventual consistency with extremely resilient availability. We'll cover consistency tuning in great detail in *Chapter 10, How Cassandra Distributes Data*.

Discretely writable collections

While it's useful for records to be internally structured into discrete fields, a given property of a record isn't always a single value like a string or an integer. One simple way to handle fields that contain collections of values is to serialize them using a format like JSON, and then save the serialized collection into a text field. However, in order to update collections stored in this way, the serialized data must be read from the database, decoded, modified, and then written back to the database in its entirety. If two clients try to perform this kind of modification to the same record concurrently, one of the updates will be overwritten by the other.

For this reason, many databases offer built-in collection structures that can be discretely updated: values can be added to, and removed from collections, without reading and rewriting the entire collection. Cassandra is no exception, offering list, set, and map collections, and supporting operations like "append the number 3 to the end of this list". Neither the client nor Cassandra itself needs to read the current state of the collection in order to update it, meaning collection updates are also blazingly efficient.

Relational joins

In real-world applications, different pieces of data relate to each other in a variety of ways. Relational databases allow us to perform queries that make these relationships explicit, for instance, to retrieve a set of events whose location is in the state of New York (this is assuming events and locations are different record types). Cassandra, however, is not a relational database, and does not support anything like joins. Instead, applications using Cassandra typically denormalize data and make clever use of clustering in order to perform the sorts of data access that would use a join in a relational database.

For data sets that aren't already denormalized, applications can also perform client-side joins, which mimic the behavior of a relational database by performing multiple queries and joining the results at the application level. Client-side joins are less efficient than reading data that has been denormalized in advance, but offer more flexibility. We'll cover both of these approaches in *Chapter 6, Denormalizing Data for Maximum Performance*.

MapReduce

MapReduce is a technique for performing aggregate processing on large amounts of data in parallel; it's a particularly common technique in data analytics applications. Cassandra does not offer built-in MapReduce capabilities, but it can be integrated with Hadoop in order to perform MapReduce operations across Cassandra data sets, or Spark for real-time data analysis. The DataStax Enterprise product provides integration with both of these tools out-of-the-box.

Comparing Cassandra to the alternatives

Now that you've got an in-depth understanding of the feature set that Cassandra offers, it's time to figure out which features are most important to you, and which database is the best fit. The following table lists a handful of commonly used databases, and key features that they do or don't have:

Feature	Cassandra	PostgreSQL	MongoDB	Redis	Riak
Structured records	Yes	Yes	Yes	Limited	No
Secondary indexes	Yes	Yes	Yes	No	Yes
Discretely writable collections	Yes	Yes	Yes	Yes	No
Relational joins	No	Yes	No	No	No
Built-in MapReduce	No	No	Yes	No	Yes
Fast result ordering	Yes	Yes	Yes	Yes	No
Immediate consistency	Configurable at query level	Yes	Yes	Yes	Configurable at cluster level
Transparent sharding	Yes	No	Yes	No	Yes
No single point of failure	Yes	No	No	No	Yes
High throughput writes	Yes	No	No	Yes	Yes

As you can see, Cassandra offers a unique combination of scalability, availability, and a rich set of features for modeling and accessing data.

Installing Cassandra

Now that you're acquainted with Cassandra's substantial powers, you're no doubt chomping at the bit to try it out. Happily, Cassandra is free, open source, and very easy to get running.

Cassandra Server

Cassandra releases include the core server, the nodetool administration command-line interface, and a development shell (cqlsh and the old cassandra-cli).
The latest stable release of Apache Cassandra is 2.1.2 (released on 2014-11-10). *If you're just starting out, download this one.*

Apache provides binary tarballs and Debian packages:
- apache-cassandra-2.1.2-bin.tar.gz [PGP] [MD5] [SHA1]
- Debian installation instructions

Third Party Distributions (not endorsed by Apache)

- DataStax Community is available in Linux rpm, deb, and tar packages, a Windows MSI installer, and a Mac OS X binary.

CQL

The current major version of the Cassandra Query Language is CQL version 3. The CQL3 documentation for the current version of Cassandra (2.0) is available here. The documentation for CQL3 in Cassandra 1.2 is here.

New users to Cassandra should be sure to check out the Getting Started guide.

Previous and Archived Cassandra Server Releases

Previous stable branches of Cassandra continue to see periodic maintenance for some time after a new major release is made. The lastest release on the 2.0 branch is 2.0.11 (released on 2014-09-24).

- apache-cassandra-2.0.11-bin.tar.gz [PGP] [MD5] [SHA1]

Installing on Mac OS X

First, we need to make sure that we have an up-to-date installation of the Java Runtime Environment. Open the Terminal application, and type the following into the command prompt:

```
$ java -version
```

You will see an output that looks something like the following:

```
java version "1.8.0_25"
Java(TM) SE Runtime Environment (build 1.8.0_25-b17)
Java HotSpot(TM) 64-Bit Server VM (build 25.25-b02, mixed mode)
```

Pay particular attention to the java version listed: if it's lower than 1.7.0_25, you'll need to install a new version. If you have an older version of Java or if Java isn't installed at all, head to `https://www.java.com/en/download/mac_download.jsp` and follow the download instructions on the page.

You'll need to set up your environment so that Cassandra knows where to find the latest version of Java. To do this, set up your JAVA_HOME environment variable to the install location, and your PATH to include the executable in your new Java installation as follows:

```
$ export JAVA_HOME="/Library/Internet Plug-
Ins/JavaAppletPlugin.plugin/Contents/Home"
$ export PATH="$JAVA_HOME/bin":$PATH
```

Downloading the example code

You can download the example code files from your account at `http://www.packtpub.com` for all the Packt Publishing books you have purchased. If you purchased this book elsewhere, you can visit `http://www.packtpub.com/support` and register to have the files e-mailed directly to you.

You should put these two lines at the bottom of your `.bashrc` file to ensure that things still work when you open a new terminal.

The installation instructions given earlier assume that you're using the latest version of Mac OS X (at the time of writing this, 10.10 Yosemite). If you're running a different version of OS X, installing Java might require different steps. Check out `https://www.java.com/en/download/faq/java_mac.xml` for detailed installation information.

Once you've got the right version of Java, you're ready to install Cassandra. It's very easy to install Cassandra using Homebrew; simply type the following:

```
$ brew install cassandra
$ pip install cassandra-driver cql
$ cassandra
```

Here's what we just did:

- Installed Cassandra using the Homebrew package manager
- Installed the CQL shell and its dependency, the Python Cassandra driver
- Started the Cassandra server

Installing on Ubuntu

First, we need to make sure that we have an up-to-date installation of the Java Runtime Environment. Open the Terminal application, and type the following into the command prompt:

```
$ java -version
```

You will see an output that looks similar to the following:

```
java version "1.7.0_65"

OpenJDK Runtime Environment (IcedTea 2.5.3) (7u71-2.5.3-
0ubuntu0.14.04.1)

OpenJDK 64-bit Server VM (build 24.65-b04, mixed mode)
```

> Pay particular attention to the java version listed: it should start with 1.7. If you have an older version of Java, or if Java isn't installed at all, you can install the correct version using the following command:
> ```
> $ sudo apt-get install openjdk-7-jre-headless
> ```

Once you've got the right version of Java, you're ready to install Cassandra. First, you need to add Apache's Debian repositories to your sources list. Add the following lines to the file /etc/apt/sources.list:

```
deb http://www.apache.org/dist/cassandra/debian 21x main
deb-src http://www.apache.org/dist/cassandra/debian 21x main
```

In the Terminal application, type the following into the command prompt:

```
$ gpg --keyserver pgp.mit.edu --recv-keys F758CE318D77295D
$ gpg --export --armor F758CE318D77295D | sudo apt-key add -
$ gpg --keyserver pgp.mit.edu --recv-keys 2B5C1B00
$ gpg --export --armor 2B5C1B00 | sudo apt-key add -
$ gpg --keyserver pgp.mit.edu --recv-keys 0353B12C
$ gpg --export --armor 0353B12C | sudo apt-key add -
$ sudo apt-get update
$ sudo apt-get install cassandra
$ cassandra
```

Here's what we just did:

- Added the Apache repositories for Cassandra 2.1 to our sources list
- Added the public keys for the Apache repo to our system and updated our repository cache
- Installed Cassandra
- Started the Cassandra server

Installing on Windows

The easiest way to install Cassandra on Windows is to use the DataStax Community Edition. DataStax is a company that provides enterprise-level support for Cassandra; they also release Cassandra packages at both free and paid tiers. DataStax Community Edition is free, and does not differ from the Apache package in any meaningful way.

DataStax offers a graphical installer for Cassandra on Windows, which is available for download at planetcassandra.org/cassandra. On this page, locate **Windows Server 2008/Windows 7 or Later (32-Bit)** from the Operating System menu (you might also want to look for 64-bit if you run a 64-bit version of Windows), and choose MSI Installer (2.x) from the version columns.

Download and run the MSI file, and follow the instructions, accepting the defaults:

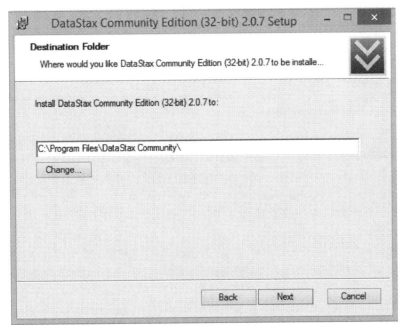

Once the installer completes this task, you should have an installation of Cassandra running on your machine.

Bootstrapping the project

Throughout the remainder of this book, we will build an application called MyStatus, which allows users to post status updates for their friends to read. In each chapter, we'll add new functionality to the MyStatus application; each new feature will also introduce a new aspect of Cassandra.

CQL – the Cassandra Query Language

Since this is a book about Cassandra and not targeted to users of any particular programming language or application framework, we will focus entirely on the database interactions that MyStatus will require. Code examples will be in **Cassandra Query Language (CQL)**. Specifically, we'll use version 3.1.1 of CQL, which is available in Cassandra 2.0.6 and later versions.

As the name implies, CQL is heavily inspired by SQL; in fact, many CQL statements are equally valid SQL statements. However, CQL and SQL are not interchangeable. CQL lacks a grammar for relational features such as JOIN statements, which are not possible in Cassandra. Conversely, CQL is not a subset of SQL; constructs for retrieving the update time of a given column, or performing an update in a lightweight transaction, which are available in CQL, do not have an SQL equivalent.

> Throughout this book, you'll learn the important constructs of CQL. Once you've completed reading this book, I recommend you to turn to the DataStax CQL documentation for additional reference. This documentation is available at `http://www.datastax.com/documentation/cql/3.1`.

Interacting with Cassandra

Most common programming languages have drivers for interacting with Cassandra. When selecting a driver, you should look for libraries that support the **CQL binary protocol**, which is the latest and most efficient way to communicate with Cassandra.

> The CQL binary protocol is a relatively new introduction; older versions of Cassandra used the Thrift protocol as a transport layer. Although Cassandra continues to support Thrift, avoid Thrift-based drivers, as they are less performant than the binary protocol.

Here are CQL binary drivers available for some popular programming languages:

Language	Driver	Available at
Java	DataStax Java Driver	`github.com/datastax/java-driver`
Python	DataStax Python Driver	`github.com/datastax/python-driver`
Ruby	DataStax Ruby Driver	`github.com/datastax/ruby-driver`
C++	DataStax C++ Driver	`github.com/datastax/cpp-driver`
C#	DataStax C# Driver	`github.com/datastax/csharp-driver`
JavaScript (Node.js)	node-cassandra-cql	`github.com/jorgebay/node-cassandra-cql`
PHP	phpbinarycql	`github.com/rmcfrazier/phpbinarycql`

While you will likely use one of these drivers in your applications, to try out the code examples in this book, you can simply use the **cqlsh** tool, which is a command-line interface for executing CQL queries and viewing the results. To start cqlsh on OS X or Linux, simply type `cqlsh` into your command line; you should see something like this:

```
$ cqlsh
Connected to Test Cluster at localhost:9160.
[cqlsh 4.1.1 | Cassandra 2.0.7 | CQL spec 3.1.1 | Thrift protocol
19.39.0]
Use HELP for help.
cqlsh>
```

On Windows, you can start cqlsh by finding the Cassandra CQL Shell application in the DataStax Community Edition group in your applications. Once you open it, you should see the same output we just saw.

Creating a keyspace

A **keyspace** is a collection of related tables, equivalent to a database in a relational system. To create the keyspace for our MyStatus application, issue the following statement in the CQL shell:

```
CREATE KEYSPACE "my_status"
WITH REPLICATION = {
  'class': 'SimpleStrategy', 'replication_factor': 1
};
```

Here we created a keyspace called `my_status`, which we will use for the remainder of this book. When we create a keyspace, we have to specify replication options. Cassandra provides several strategies for managing replication of data; `SimpleStrategy` is the best strategy as long as your Cassandra deployment does not span multiple data centers. The `replication_factor` value tells Cassandra how many copies of each piece of data are to be kept in the cluster; since we are only running a single instance of Cassandra, there is no point in keeping more than one copy of the data. In a production deployment, you would certainly want a higher replication factor; 3 is a good place to start.

A few things at this point are worth noting about CQL's syntax:

- It's syntactically very similar to SQL; as we further explore CQL, the impression of similarity will not diminish.
- Double quotes are used for identifiers such as keyspace, table, and column names. As in SQL, quoting identifier names is usually optional, unless the identifier is a keyword or contains a space or another character that will trip up the parser.
- Single quotes are used for string literals; the key-value structure we use for replication is a **map** literal, which is syntactically similar to an object literal in JSON.
- As in SQL, CQL statements in the CQL shell must terminate with a semicolon.

Selecting a keyspace

Once you've created a keyspace, you would want to use it. In order to do this, employ the USE command:

```
USE "my_status";
```

This tells Cassandra that all future commands will implicitly refer to tables inside the `my_status` keyspace. If you close the CQL shell and reopen it, you'll need to reissue this command.

Summary

In this chapter, you explored the reasons to choose Cassandra from among the many databases available, and having determined that Cassandra is a great choice, you installed it on your development machine.

You had your first taste of the Cassandra Query Language when you issued your first command via the CQL shell in order to create a keyspace. You're now poised to begin working with Cassandra in earnest.

In the next chapter, we'll begin building the MyStatus application, starting out with a simple table to model users. We'll cover a lot more CQL commands, and before you know it, you'll be reading and writing data like a pro.

2
The First Table

For the next few chapters, we'll be building the persistence layer for an application called MyStatus, which allows its users to post status updates for their friends to see. In this chapter, we'll make the first table of that persistence layer, which will store user profiles.

In *Chapter 1, Getting Up and Running with Cassandra*, you set up a local Cassandra installation and created a keyspace, so we're prepared to hit the ground running. This chapter will provide a whirlwind tour of many core Cassandra concepts such as:

- How Cassandra tables are structured
- Cassandra's type system
- How to create a table
- How to insert rows into a table
- How to retrieve rows by primary key
- How to query for all rows in a table
- How to page through large result sets

Creating the users table

Our first table will store basic user account information: username, email, and password. To create the table, fire up the CQL shell (don't forget to use the USE "my_status"; statement if you are starting a fresh session) and enter the following CQL statement:

```
CREATE TABLE "users" (
  "username" text PRIMARY KEY,
  "email" text,
  "encrypted_password" blob
);
```

In the above statement, we created a new table called users, which has three columns: username and email, which are text columns, and encrypted_password, which has the type blob. The username column acts as the primary key for the table.

Structuring of tables

Cassandra structures tables in rows and columns, just like a relational database. Also like a relational database, the columns available to a table are defined in advance. New columns cannot be added on-the-fly when inserting data, although it's possible to update an existing table's schema.

Every table defines one or more columns to act as the **primary key**; each row is uniquely identified by the value(s) in its primary key column(s), and those columns cannot be left blank in any row. Cassandra does not offer auto-incrementing primary keys; each row, when created, must be explicitly assigned a primary key by the client. One good way to structure the primary key is to use a **natural key**, which is a value that is fundamentally unique for each row you want to store. That's what we do in the users table by making username, a naturally unique identifier, the primary key.

Before we continue, I should emphasize the importance of using the latest version of Cassandra. Cassandra and CQL have changed substantially over the past few years, and the current CQL version 3.1 is not backward compatible with older versions of the language. The examples in this book target the latest Cassandra version (at the time of writing, 2.1.2); make sure you're running this version or a newer one.

Table and column options

The table creation statement given in the previous section is very simple: little more than a list of column names and their respective types. In contrast, an SQL CREATE TABLE statement will often look quite complex, with multiple options set on each column, such as defaults, constraints, and nullability flags. What sorts of bells and whistles can we add to a CQL table?

At the table level, Cassandra does have quite a few configuration options, known as **table properties**. These properties allow you to tune a wide range of under-the-hood aspects of the table, such as caching, compression, garbage collection, and read repair. Table properties do not, however, bear on the table's behavior from the application's standpoint. For this reason, we won't go into detail about them in this book.

Columns, on the other hand, have very few knobs to turn. In fact, outside of the type, each column is pretty much the same. There are several column options you may be accustomed to from SQL databases that don't carry over to Cassandra, such as the following:

- Cassandra doesn't have a concept of NULL values; columns either have data, or they don't. Primary key columns are always required to have a value; non-key columns are always optional. You will see the word null appear in the cqlsh output, but that simply means there is no data in this column, and should not be confused with the concept of NULL in a relational database.

- Cassandra doesn't support default values for columns. If a row is inserted without a value for a certain column, that column just doesn't have a value.

- Cassandra doesn't provide data validations like length limits or other more complex column constraints. As long as a value is of the right type for the column you're putting it in, it's valid.

Happily, most modern applications do not need to rely on the sorts of constraints listed above; domain modeling libraries and object mappers typically allow you to easily apply these constraints at the application level.

The type system

Each column defined for a table has a defined type. In the users table, we used the text and blob types, but those are only two of the many types built in to Cassandra.

Strings

Cassandra has two types that store string data:

- The `text` type stores UTF-8 encoded strings. It's also aliased as the `varchar` type; you can use these interchangeably.

- The `ascii` type stores strings of ASCII characters (bytes 0–127).

Neither of the above string types has a limit on the length of strings that can be stored. In CQL, string literals are surrounded in single quotes, `'like this'`.

Integers

Cassandra has three types that store integers:

- The `int` type stores 32-bit integers, which can store values ranging from approximately -2.1 billion to 2.1 billion

- The `bigint` type stores 64-bit integers, which can store values from about -9 quintillion to 9 quintillion

- The `varint` type stores integers of arbitrary size; it has no minimum or maximum value

All integer types are signed, meaning they can hold positive or negative integers. There are no unsigned numeric types in Cassandra. Integer literals in CQL, like in most languages, consist of an optional minus sign followed by one or more digits, such as `3549`.

Floating point and decimal numbers

Cassandra has three types that store non-integer numbers:

- The `float` type stores 32-bit floating point numbers.

- The `double` type stores 64-bit floating point numbers.

- The `decimal` type stores variable-precision decimal numbers, with no upper bound on size. Unlike a floating point number, a variable-precision decimal will never suffer from base 10 rounding errors in the fractional part of the number.

Like the integer types, the floating point and decimal types are always signed. In CQL, floating point and decimal literals are represented by an optional minus sign, followed by a series of digits, followed by a period, followed by another series of digits, such as `7152.6846`. You can also write them using exponential notation, like `9.021e14`.

Dates and times

Dates and times can be stored using the `timestamp` type, which holds date/time data at millisecond precision. Timestamp literals are enclosed in single quotation marks like string literals, and take the format `'yyyy-mm-dd HH:mm:ssZ'`, for example `'2014-05-18 15:49:31-0400'`. There is no type that stores dates without times, although the time portion of a timestamp literal can be omitted, which defaults the time to midnight in the given time zone. To represent timestamps with millisecond precision, you may also use a numeric literal with the number of milliseconds since midnight UTC on January 1, 1970, for instance, `1400442761830`.

UUIDs

Cassandra has two types that store universally unique identifiers:

- The `uuid` type stores Version 1 and Version 4 UUIDs
- The `timeuuid` type stores Version 1 UUIDs, and has special functionality for conversion between UUIDs and timestamps

> A **UUID**, which is short for **universally unique identifier**, is essentially a very large number generated in a specific way, designed to guarantee that the same UUID will never be generated anywhere in the world at any time. Version 1 UUIDs are generated using a high-precision timestamp and the generating computer's MAC address; the timestamp can be extracted from the UUID. Version 4 UUIDs use random or pseudorandom numbers.

CQL uses the canonical representation of UUIDs, which is a sequence of hexadecimal digits broken up in specific places by dashes, in the form *8-4-4-4-12*. A CQL UUID literal is not surrounded by quotation marks or any other delimiter; for example, `550e8400-e29b-41d4-a716-446655440000` is a valid UUID literal. Most languages have UUID libraries available that will generate UUIDs and output them in the canonical format.

Booleans

The `boolean` type stores simple true/false values. A boolean literal is either `true` or `false`, with no surrounding quotation marks.

Blobs

The `blob` type stores unstructured binary data. Blobs are a good choice for images, audio, and encrypted data. In CQL, the blob literal is a sequence of hexadecimal digits, prefixed with `0x`, for instance, `0x1d4375023013dba2d5f9a`. Blob literals are not surrounded by quotation marks.

The purpose of types

Now that you know the full range of types available in Cassandra, you may be wondering what purpose those types serve. In fact, the type system in Cassandra plays a few roles:

- Types are used for input validation. If you attempt to put a string value in an integer column, for instance, Cassandra will return an error.

- Type information is made available to client libraries; most adapters will return the results of queries with values represented using the appropriate data type for the language.

- In some scenarios, rows can be ordered by the value of a certain column. In that case, the type of the column determines the order of values in the column. For instance, in a `text` column, `'2'` is larger than `'10'`, but in an `int` column, `10` is larger than `2`.

 You can find a full list of all the CQL data types in the DataStax CQL documentation, at http://www.datastax.com/documentation/cql/3.1/cql/cql_reference/cql_data_types_c.html.

Now that we've got a firm grasp of how to create a Cassandra table and what options are available to us when creating columns, it's time to put the table to use.

Inserting data

For our status sharing application, the first thing we'll want any user to do is to create an account. We'll ask them to choose a username, enter their email, and pick a password; our business logic will be responsible for ensuring that the entries are valid and for encrypting the password appropriately. At that point, we'll be ready to insert the account information as a new row in the `users` table:

```
INSERT INTO "users"
("username", "email", "encrypted_password")
VALUES (
  'alice',
  'alice@gmail.com',
  0x8914977ed729792e403da53024c6069a9158b8c4
);
```

In the previous statement, which should be familiar to anyone who has used an SQL database, we provide the following information:

- We want to add a row to the users table
- We'll be adding data to three columns in that row: username, email, and encrypted_password
- Finally, we provide the values to insert into those columns, in the same order that the column names were listed previously

Does whitespace matter?

CQL is agnostic about whitespace. The query as written uses a lot of newlines for readability, but the query would be just as valid written on a single line.

Type the preceding CQL statement into your Cassandra shell (don't forget about the semicolon at the end!) Did it work? At first glance, it may be hard to tell.

Writing data does not yield feedback

If the INSERT statement worked, you should not see any response in the shell whatsoever; it should simply provide you with a new command prompt. This is not simply a quirk of the CQL shell, but a fundamental fact about writing data in Cassandra: writing data does not normally result in any information about the operation from the database, other than an error message if the write failed. Most client libraries will return a null value, or the equivalent in the language in question, when you perform a successful write query.

This may come as a surprise if you're used to working with an SQL database, which will typically give you detailed feedback when you write data, such as returning the primary key of the new row or simply telling you how many rows were affected by the operation.

We'll see some exceptions to this rule when we explore lightweight transactions in *Chapter 7, Expanding Your Data Model*.

Partial inserts

You are not, of course, required to provide values for all columns — other than primary key columns — when inserting a row. If, for instance, we decide to allow users to register without providing an email, we may issue this perfectly valid query:

```
INSERT INTO "users"
("username", "encrypted_password")
VALUES (
  'bob',
  0x10920941a69549d33aaee6116ed1f47e19b8e713
);
```

In the above query, we only insert values for the `username` and `encrypted_password` fields; the row will have no value in the `email` field.

Do empty columns take up space?

Only columns with values take up storage space in Cassandra. This is in contrast to relational databases in which every row has space allocated for every column, whether or not that column has a value. So, there's little downside to defining columns in Cassandra that you expect to rarely populate; they'll only take up space where they have values.

For a full reference on the `INSERT` statement in CQL, consult the DataStax CQL documentation at `http://www.datastax.com/documentation/cql/3.1/cql/cql_reference/insert_r.html`. We haven't explored all of the possibilities for `INSERT` statements yet, but we'll cover many of them in future chapters.

Selecting data

We now know how to retrieve data from the database, but that isn't much good unless we can get it back again. Let's say we now want to build an account settings page for MyStatus; we've got the user's username stored in a persistent session, but we will retrieve the other profile fields from the database to display in the settings form:

```
SELECT * FROM "users"
WHERE "username" = 'alice';
```

This query tells Cassandra we want to retrieve the rows where the value for username (the primary key) is alice. The * wildcard simply says we would like all the columns in that row, saving us from having to type them all out. You'll see the rows we requested nicely formatted in the CQL shell as follows:

```
username | email            | encrypted_password
---------+------------------+---------------------------------------------
   alice | alice@gmail.com  | 0x8914977ed729792e403da53024c6069a9158b8c4
```

In other scenarios, we don't need all the columns. When a user tries to log in to MyStatus, we want to retrieve their password and compare it to the one the user provided us with, but we don't care about the email. Avoiding unnecessary columns reduces the amount of data that needs to be transferred between Cassandra and our application, thus making the queries faster. Instead of using a wildcard (*), we can instead type a list of columns we are interested in:

```
SELECT "username", "encrypted_password" FROM "users"
WHERE "username" = 'alice';
```

You'll see that in the results, the email column no longer appears, since it wasn't in the list of columns that we specified:

```
username | encrypted_password
---------+---------------------------------------------
   alice | 0x8914977ed729792e403da53024c6069a9158b8c4
```

Missing rows

What happens when we ask for a primary key that doesn't exist? Let's try it out:

```
SELECT * FROM "users"
WHERE "username" = 'bogus';
```

You'll see that Cassandra simply returns no results; it is not an error to try to retrieve a primary key that does not exist.

Selecting more than one row

Let's say we'd like to build an administrative interface that allows employees to access data for several users in one screen. We could, of course, simply perform a query for each username specified, but Cassandra gives us a more efficient way to do this:

```
SELECT * FROM "users"
WHERE "username" IN ('alice', 'bob');
```

This query will return two rows: one with the alice primary key and the other with the bob primary key:

```
username | email             | encrypted_password
---------+-------------------+--------------------------------------------
   alice | alice@gmail.com   | 0x8914977ed729792e403da53024c6069a9158b8c4
     bob |            null   | 0x10920941a69549d33aaee6116ed1f47e19b8e713
```

Note that while this will be faster than performing two queries, it does require Cassandra to perform two seeks for the two rows, so querying for the additional row comes at some cost.

Retrieving all the rows

Perhaps we'd like to expand our administrative interface to show a list of all the users who've signed up for MyStatus. To do this, we'll want to simply ask Cassandra for all the user records:

```
SELECT * FROM "users";
```

Since we omitted the WHERE portion of the query, Cassandra will simply return all rows in the users table:

```
username | email             | encrypted_password
---------+-------------------+--------------------------------------------
     bob |            null   | 0x10920941a69549d33aaee6116ed1f47e19b8e713
   alice | alice@gmail.com   | 0x8914977ed729792e403da53024c6069a9158b8c4
```

If you've been following along with the examples, you'll see that the latest query returns both rows that we've inserted into the database. However, had we inserted over 10,000 rows, we'd notice that *only 10,000 rows would be returned*. This is a limit built into cqlsh; using a language driver, you can retrieve result sets of arbitrary size.

You'll also notice that bob is returned before alice. Clearly, rows are not returned in lexical order of their primary key. As it happens, in a table with a single primary key column, row ordering is deterministic but not defined from a client standpoint. Deterministic ordering is enough, however, to allow us to page through very large result sets.

 You'll find a full reference for the SELECT statement in the DataStax CQL documentation at http://www.datastax. com/documentation/cql/3.1/cql/cql_reference/ select_r.html.

Paginating through results

In most situations, we will want to avoid displaying an arbitrary number of results to a user in a single response. Instead, we will display a fixed number of results, and give the user an interface to load additional pages of the same size. In an SQL database, pages are typically specified using the OFFSET keyword, but CQL does not have this capability. Instead, we'll use the natural ordering of primary keys to put a lower bound on the key values displayed on a given page.

Let's insert a couple more rows into our table to get a sense of how pagination works using the following INSERT statements:

```
INSERT INTO "users"
("username", "email", "encrypted_password")
VALUES (
  'carol',
  'carol@gmail.com',
  0xed3d8299b191b59b7008759a104c10af3db6e63a
);
```

```
INSERT INTO "users"
("username", "email", "encrypted_password")
VALUES (
  'dave',
  'dave@gmail.com',
  0x6d1d90d92bbab0012270536f286d243729690a5b
);
```

Now that we've got four users in our system, we can paginate over them. To save ourselves the trouble of adding numerous additional entries by hand, we'll just use a page size of two to demonstrate the process. We'll start by retrieving the first page as follows:

```
SELECT * FROM users
LIMIT 2;
```

The LIMIT part of the query simply tells Cassandra to return no more than two results:

```
 username | email           | encrypted_password
----------+-----------------+------------------------------------------
      bob |            null | 0x10920941a69549d33aaee6116ed1f47e19b8e713
     dave | dave@gmail.com  | 0x6d1d90d92bbab0012270536f286d243729690a5b
```

Now that we have our first page, we want to get the second. It would be nice if we could simply ask for the next primary key in order after dave using the following SELECT statement:

```
SELECT * FROM "users"
WHERE "username" > 'dave'
LIMIT 2;
```

Unfortunately, this will give an error:

```
Bad Request: Only EQ and IN relation are supported on the partition key
(unless you use the token() function)
```

The message is a bit cryptic, and we don't know what exactly a *partition key* is yet, but it does contain a helpful hint: we can use the token() function to do what we want. The reason that our attempted query doesn't work is that, as we noticed before, the primary keys in our users table are not stored in lexical order; Cassandra can only return rows in the order in which they are stored.

The actual ordering is determined by the **token** of the primary key; the way the token is calculated is opaque to us, but Cassandra lets us use the `token()` function to retrieve the token for a given value:

```
SELECT "username", token("username")
FROM "users";
```

Now we can see why the rows are returned in the order they are; they ascend by token:

```
username | token(username)
----------+---------------------
     bob | -5396685590450884643
    dave | -4493667438046306776
   carol | -3169904368870211108
   alice |  5699955792253506986
```

Armed with this function, we can retrieve the next page of results as follows:

```
SELECT * FROM "users"
WHERE token("username") > token('dave')
LIMIT 2;
```

And just as we'd hoped, the next two rows are returned as shown below:

```
username | email           | encrypted_password
----------+-----------------+-------------------------------------------
   carol | carol@gmail.com | 0xed3d8299b191b59b7008759a104c10af3db6e63a
   alice | alice@gmail.com | 0x8914977ed729792e403da53024c6069a9158b8c4
```

Using this technique, we can paginate over arbitrarily large tables using multiple queries.

 It bears emphasizing that retrieving large result sets from tables structured like `users` is a relatively expensive operation for Cassandra. In *Chapter 3, Organizing Related Data*, we'll begin to develop a more advanced table structure that will allow us to retrieve batches of rows very efficiently.

Developing a mental model for Cassandra

Any time you learn a new tool, you will naturally begin to develop a mental model for how that tool works. So far, our model for how Cassandra tables works is fairly simple; we will expand upon it throughout the book.

For now, we can think of a Cassandra table as a collection of keys, each of which points to a row. Each row contains data in some subset of its columns.

We also know that, at least in the users table, rows are stored non-contiguously; accessing each row requires Cassandra to seek a different place in storage. So we may imagine the current state of our users table to look something like this:

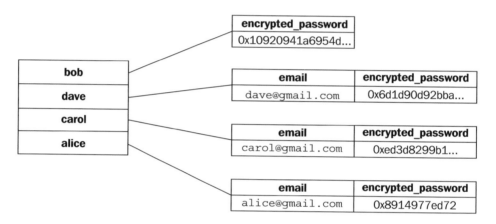

Essentially, we can think of our Cassandra table as a key-value store, where each value consists of one or more predefined columns containing data of a predefined type.

Summary

In this chapter, you're off to a running start with your MyStatus application, having created our first table, inserted data into it, and learned a few ways to retrieve that data. We've developed a model for how Cassandra tables are structured, and taken a deep dive into Cassandra's type system.

We now have experience with the INSERT and SELECT statements, two of the core operations in CQL. You've learned that primary keys, at least the ones we've seen so far, determine the order in which rows are returned, but that ordering is opaque to an application. We can only rely on it being consistent. We've seen some of the limitations of simple tables in Cassandra, and we know that the types of multi-row retrieval we know so far aren't the most efficient.

In *Chapter 3, Organizing Related Data*, we'll introduce a new way to structure a table that lets us overcome many of the limitations we encountered here.

3

Organizing Related Data

In *Chapter 2, The First Table*, we created our first table, which stores user accounts. We discussed how to insert data into the table, and how to retrieve it. However, we also encountered several significant limitations in the tasks we can perform with the table we created.

In this chapter, we'll introduce the concept of compound primary keys, which are simply primary keys comprising more than one column. Although this might at first glance seem like a trivial addition to our understanding of Cassandra tables, a table with compound primary keys, in fact, is a considerably richer data structure that opens up substantial new data access patterns.

Our introduction to compound primary keys will help us to build a table that stores a timeline of users' status updates. In this chapter, we'll focus on defining the table and understanding how it works; *Chapter 4, Beyond Key-Value Lookup*, will introduce new patterns to query compound primary key tables.

We'll explore two different approaches to designing schemas with compound primary keys. In the first approach, the primary key encodes a parent-child relationship implicitly: In this case, a user's status updates are children of the user record itself. We'll also look at an alternative schema design using static columns; this design allows us to store information about users and their status updates in a single table, without duplication. This makes the relationship between users and status updates explicit in the structure of the table.

By the end of this chapter, you'll know:

- How to create a table with a compound primary key
- The difference between a partition key and a clustering column
- How Cassandra organizes data in a compound key table
- When to use UUIDs for primary key components
- How to use static columns to associate data with partition keys
- When static columns are useful for schema design

A table for status updates

In the MyStatus application, we'll begin by creating a timeline of status updates for each user. Users can view their friends' status updates by accessing the timeline of the friend in question.

The user timeline requires a new level of organization that we didn't see in the `users` table that we created in the previous chapter. Specifically, we have two requirements:

- Rows (individual status updates) should be logically grouped by a certain property (the user who created the update)
- Rows should be accessible in sorted order (in this case, by creation date)

Fortunately, compound primary keys provide exactly these qualities.

Creating a table with a compound primary key

The syntax for creating tables with compound primary keys is a bit different from the single-column primary key syntax we saw in the previous chapter. We create a `user_status_updates` table with a compound primary key, as follows:

```
CREATE TABLE "user_status_updates" (
  "username" text,
  "id" timeuuid,
  "body" text,
  PRIMARY KEY ("username", "id")
);
```

In particular, instead of appending a single column definition with a PRIMARY KEY modifier, we make a separate PRIMARY KEY declaration at the end of the list of columns, which itself specifies the two columns that make up the **compound primary key**. While this is the only way to declare a compound primary key, it's also a perfectly valid way to declare a single-column primary key. So, our users table from the previous chapter could have been declared like this:

```
CREATE TABLE "users" (
  "username" text,
  "email" text,
  "encrypted_password" blob,
  PRIMARY KEY ("username")
);
```

Here, we simply move the PRIMARY KEY declaration to the end of the column list. It's a little less concise, but it has the exact same effect as a single column definition with a PRIMARY KEY modifier.

The structure of the status updates table

The most notable aspect of user_status_updates is that it has two columns for a primary key. This means that each row is identified uniquely by the combination of its username and id columns. It also means that every row must have a value in both of these columns.

In addition to this, our user_status_updates table is the first time we've seen a timeuuid column in the wild. As you can recollect from the previous chapter, a UUID is essentially a very large number that is generated using an algorithm that guarantees that the identifier is unique across time and space.

You will also recollect that Cassandra does not have the ability to generate auto-incrementing sequences for use in primary keys, as this would require an unacceptably high level of coordination between nodes in the cluster. In the users table, we used a natural key; we want each user to have a unique username, so the username column makes a perfectly good unique identifier for rows.

In the case of a **status update**, however, there is no obvious natural key. The only user-generated data associated with a status update is the body, but a free text field doesn't make a very good primary key, and anyway there's no guarantee that status update bodies will be unique. This is where UUIDs come in handy. Since they're guaranteed to be unique, we can use them as a **surrogate key**—a unique identifier that isn't derived from the data in the row. Auto-incrementing primary keys in relational databases are also surrogate keys.

UUIDs and timestamps

While there are several algorithms that can be used to generate UUIDs, the **Version 1 UUID** has an additional useful property: part of the UUID encodes the timestamp at which the UUID is generated. This timestamp can be extracted from the full UUID, meaning that it's possible to know exactly when any Version 1 UUID was generated.

Cassandra's `timeuuid` type lets us capitalize on that property. Cassandra is aware of the structure of a `timeuuid`, and is able to both convert timestamps into UUIDs and to extract the creation timestamp from a UUID. As we'll soon see, Cassandra can also sort our rows by their creation time using the timestamps encoded in the UUIDs.

Working with status updates

Now that we've got our status updates table ready, let's create our first status update:

```
INSERT INTO "user_status_updates"
("username", "id", "body")
VALUES (
  'alice',
  76e7a4d0-e796-11e3-90ce-5f98e903bf02,
  'Learning Cassandra!'
);
```

This will look pretty familiar; we specify the table we want to insert data into, the list of columns we're going to provide data for, and the values for these columns in the given order.

Let's give bob a status update too, by inserting the following row in the user_status_updates table:

```
INSERT INTO "user_status_updates"
("username", "id", "body")
VALUES (
  'bob',
  97719c50-e797-11e3-90ce-5f98e903bf02,
  'Eating a tasty sandwich.'
);
```

Now we have two rows, each identified by the combination of the username and id columns. Let's take a look at the contents of our table using the following SELECT statement:

```
SELECT * FROM "user_status_updates";
```

We'll be able to see the two rows that we inserted, as follows:

```
username | id                                   | body
---------+--------------------------------------+----------------------------
     bob | 97719c50-e797-11e3-90ce-5f98e903bf02 | Eating a tasty sandwich.
   alice | 76e7a4d0-e796-11e3-90ce-5f98e903bf02 |        Learning Cassandra!
```

Note that, as we saw in the `users` table, the rows are not returned in lexical order of `username`; indeed, they're in the same order as the user records themselves are. Recall from the *Paginating through results* section of *Chapter 2, The First Table*, that the `username` column is ordered by an internally generated **token**, which is deterministic but meaningless from the application's perspective.

Extracting timestamps

As we previously mentioned, Cassandra has special capabilities for the `timeuuid` type, which includes extracting the timestamp that's encoded in these UUIDs. We can see this in action using the `DATEOF` function:

```
SELECT "username", "id", "body", DATEOF("id")
FROM "user_status_updates";
```

The `DATEOF` function instructs Cassandra to return a result column containing the timestamp at which the given column's UUID value was created. We now have access to information encoded in the `id` column that was previously obscure:

```
username | id                                   | body                     | DATEOF(id)
---------+--------------------------------------+--------------------------+-------------------------
     bob | 97719c50-e797-11e3-90ce-5f98e903bf02 | Eating a tasty sandwich. | 2014-05-29 21:13:21-0400
   alice | 76e7a4d0-e796-11e3-90ce-5f98e903bf02 |        Learning Cassandra! | 2014-05-29 21:05:17-0400
```

If you're following along in your CQL shell, you'll notice that the dates do not tell you when the rows were created, but rather when the UUIDs in the `id` column were generated. Since we're using UUIDs that I generated when writing this book, we'll see that the creation timestamp shows an older time.

While the `DATEOF` output is very readable, the timestamps only offer precision at the second level. For a more precise representation of the timestamp at which the UUIDs were generated, use the `UNIXTIMESTAMPOF` function instead:

```
SELECT "username", "id", "body", UNIXTIMESTAMPOF("id")
FROM "user_status_updates";
```

The UNIXTIMESTAMPOF function returns the timestamp represented as the number of milliseconds since January 1, 1970 at midnight UTC:

```
username | id                                     | body                   | UNIXTIMESTAMPOF(id)
---------+----------------------------------------+------------------------+--------------------
     bob | 97719c50-e797-11e3-90ce-5f98e903bf02   | Eating a tasty sandwich. |    1401412401813
   alice | 76e7a4d0-e796-11e3-90ce-5f98e903bf02   |      Learning Cassandra! |    1401411917725
```

Looking up a specific status update

In our status update application, we'll want to allow direct linking to status updates, in which case we'll need to be able to retrieve a specific status update from the table. So far, we've issued an open-ended SELECT statement to see all the rows we've inserted in user_status_updates, but we haven't seen how to return one particular row.

Since user_status_updates has a compound primary key, we need new CQL syntax to allow us to specify values for multiple columns in order to form a unique identifier. CQL provides the AND construct for this purpose:

```
SELECT * FROM "user_status_updates"
WHERE "username" = 'alice'
AND "id" = 76e7a4d0-e796-11e3-90ce-5f98e903bf02;
```

By specifying both the username and the id, we identify exactly one row:

```
username | id                                     | body
---------+----------------------------------------+---------------------
   alice | 76e7a4d0-e796-11e3-90ce-5f98e903bf02   | Learning Cassandra!
```

We now have all the tools we need to interact with the user_status_updates table in the same way as we interact with the users table; let's move on and explore some of the things that make a compound primary key table different.

Automatically generating UUIDs

In the status updates we've created so far, we've used predetermined UUIDs that were generated on May 29, 2014. For this reason, the timestamp encoded in the UUIDs doesn't really tell us anything useful about when the rows were created. However, in the general case, we would like to encode useful data in the UUID: in particular, the timestamp at which the row itself was created.

Libraries that generate Version 1 UUIDs are available for just about any programming language, but Cassandra also gives us a built-in CQL function to generate a UUID from the current time, the NOW function. Let's use that function to insert a few new status updates:

```
INSERT INTO "user_status_updates" ("username", "id", "body")
VALUES ('alice', NOW(), 'Alice Update 1');
INSERT INTO "user_status_updates" ("username", "id", "body")
VALUES ('bob', NOW(), 'Bob Update 1');
INSERT INTO "user_status_updates" ("username", "id", "body")
VALUES ('alice', NOW(), 'Alice Update 2');
INSERT INTO "user_status_updates" ("username", "id", "body")
VALUES ('bob', NOW(), 'Bob Update 2');
INSERT INTO "user_status_updates" ("username", "id", "body")
VALUES ('alice', NOW(), 'Alice Update 3');
INSERT INTO "user_status_updates" ("username", "id", "body")
VALUES ('bob', NOW(), 'Bob Update 3');
```

Instead of explicitly specifying a UUID constant for the id field, we used the NOW function to generate a new and unique UUID for each row we inserted.

While the NOW function is quite convenient, particularly in the CQL shell, it comes with one major downside. Since a CQL INSERT does not give any feedback on the results of the operation, you won't actually know what UUID the NOW function generates. Sometimes this is fine, but in many cases your application would want to perform further business logic that requires knowing the primary key of the record that was just created. In these cases, it's better to use a library to generate UUIDs at the application level and provide them explicitly as literals in the INSERT statement.

Now that we've got a handful of rows in the user_status_updates table, let's take a look at its contents:

```
SELECT "username", "id", "body", UNIXTIMESTAMPOF("id")
FROM "user_status_updates";
```

As we did previously, we'll ask Cassandra to return the millisecond-precision timestamp of the UUIDs in the table along with the data in all columns. Let's take a look at the results:

```
username | id                                   | body                  | UNIXTIMESTAMPOF(id)
---------+--------------------------------------+-----------------------+--------------------
     bob | 97719c50-e797-11e3-90ce-5f98e903bf02 | Eating a tasty sandwich. |  1401412401813
     bob | 3f9d81e0-e8f7-11e3-9211-5f98e903bf02 |         Bob Update 1 |    1401563437310
     bob | 3f9e9350-e8f7-11e3-9211-5f98e903bf02 |         Bob Update 2 |    1401563437317
     bob | 3f9f56a0-e8f7-11e3-9211-5f98e903bf02 |         Bob Update 3 |    1401563437322
   alice | 76e7a4d0-e796-11e3-90ce-5f98e903bf02 |  Learning Cassandra! |    1401411917725
   alice | 3f9b5f00-e8f7-11e3-9211-5f98e903bf02 |       Alice Update 1 |    1401563437296
   alice | 3f9df710-e8f7-11e3-9211-5f98e903bf02 |       Alice Update 2 |    1401563437313
   alice | 3f9ee170-e8f7-11e3-9211-5f98e903bf02 |       Alice Update 3 |    1401563437319
```

We notice a couple of interesting things here. First, we can see that results are grouped by username: all the status updates of bob appear first, followed by all the status updates of alice. This is despite the fact that we interleaved the status updates of alice and bob when we inserted them.

Second, within each user's status updates, the updates are returned in ascending order of the timestamp of the row's id column. Looking at the rightmost column in the results, we see that the timestamps monotonically increase for each user. This is no coincidence; the id column determines the ordering of rows in the user_status_ updates table, and since it's a timeuuid column, the timestamp encoded in the UUID determines the semantic ordering of the rows.

Anatomy of a compound primary key

At this point, it's clear that there's some nuance in the compound primary key that we're missing. Both the username column and the id column affect the order in which rows are returned; however, while the actual ordering of username is opaque, the ordering of id is meaningfully related to the information encoded in the id column.

In the lexicon of Cassandra, username is a **partition key**. A table's partition key groups rows together into logically related bundles. In the case of our MyStatus application, each user's timeline is a self-contained data structure, so partitioning the table by user is a sound strategy.

> As a general rule, you should endeavor to *only query one partition at a time* for any core data access your application does. Cassandra stores the rows in each partition together, so queries within a partition are very efficient. Queries across multiple partitions, on the other hand, are expensive and should be avoided.

We call the id column a **clustering column**. The job of a clustering column is to determine the ordering of rows within a partition. This is why we observed that within each user's status updates, the rows were returned in a strictly ascending order by timestamp of the id. This is a very useful property, since our application will want to display status updates ordered by creation time.

Is sorting by clustering column efficient?

Sorting any collection at read time is expensive for a non-trivial number of elements. Happily, Cassandra stores rows in clustering order, so when you retrieve them, it simply returns them in the order they're stored in. There's no expensive sorting operation at read time.

All of the rows that share the same primary key are stored in a contiguous structure on disk. It's within this structure that rows are sorted by their clustering column values. Because each partition is tightly bound at the storage level, there is an upper bound on the number of rows that can share the same partition key. In theory, this limit is about 2 billion total column values. For instance, if you have a table with 10 data columns, your upper bound would be 200 million rows per partition key.

For further information on data modeling using compound primary keys, the DataStax CQL documentation has a good explanation at http://www.datastax.com/documentation/cql/3.1/cql/ ddl/ddl_compound_keys_c.html

Anatomy of a single-column primary key

Now that you understand the distinction between a partition key and a clustering column, you might be wondering which role the username column plays in the users table.

As it turns out, it's a partition key. All Cassandra tables must have a partition key; clustering columns are optional. In the users table, each row is its own tiny partition; no row is grouped with any other.

Beyond two columns

We've now seen a table with two columns in its primary key: a partition key, and a clustering column. As it turns out, neither of these roles is limited to a single column. A table can define one or more partition key columns, and zero or more clustering columns.

For instance, in our status application, we might want to allow users to reply to other users' status updates. In this case, each status update would have a stream of replies; replies would be partitioned by the full primary key of the original status update, and each reply would get its own timestamped UUID:

```
CREATE TABLE "status_update_replies" (
  "status_update_username" text,
  "status_update_id" timeuuid,
  "id" timeuuid,
  "author_username" text,
  "body" text,
  PRIMARY KEY (
    ("status_update_username", "status_update_id"),
    "id"
  )
);
```

Note the extra set of parentheses around the status_update_username and status_update_id columns in the PRIMARY KEY declaration. This tells Cassandra that we want those two columns together to form the partition key. Without the extra parentheses, Cassandra assumes by default that *only the first column* in the primary key is a partition key, and the remaining columns are clustering columns.

Compound keys represent parent-child relationships

In the *What Cassandra offers, and what it doesn't* section of *Chapter 1, Getting Up and Running with Cassandra*, you learned that Cassandra is not a relational database, despite some surface similarities. Specifically, this means that Cassandra does not have a built-in concept of the relationships between data in different tables. There are no foreign key constraints and there's no JOIN clause available in the SELECT statements; in fact, there is no way to read from multiple tables in the same query. Whereas relational databases are designed to explicitly account for the relationships between data in different tables, whether they're one-to-one, one-to-many, or many-to-many. Cassandra has no built-in mechanism for describing or traversing inter-table relationships.

That being said, Cassandra's compound primary key structure provides an ample affordance for a particular kind of relationship—the **parent-child relationship**. This is a specific type of one-to-many relationship in which the "one" side plays a unique role with respect to the "many" side; we can say that the "one" is a parent or a container for the "many". We've already seen two examples of this: a user's status updates are children of the user themselves; and the comments about a status update are children of that status update.

This relationship is represented quite transparently in the compound primary key structure. The partition key acts as a reference to the parent, and the clustering column uniquely identifies the row among its siblings. This is why we used both the `status_update_username` and `status_update_id` columns for the partition key in our `status_update_replies` table; these columns together provide a full reference to the reply's parent, namely the status update to which it's a reply.

It's worth emphasizing that not every one-to-many relationship is a parent-child relationship. For instance, on a blogging platform, we'd expect a blog post to have at least a couple of many-to-one relationships, namely, an author relation and a blog relation. Only one of these can be a parent-child relationship; in the blog example, it seems natural to think of the parent of a blog post as the blog.

Our Cassandra data models can only accommodate a single parent relation for a given table because the parent relation is expressed as the partition key column(s) of the table. Not all table schemas fit this line of reasoning; sometimes a partition key is just a partition key, such as a time-series table that partitions by date. However, parent-child relationships provide a fruitful framework for Cassandra data modeling across a wide variety of applications.

Coupling parents and children using static columns

The parent-child relationships we've encoded in our schema thus far are implicit in the structure of the primary keys, but not explicit from Cassandra's standpoint. While we know that the `user_status_updates.username` column corresponds to the "parent" primary key `users.username`, Cassandra itself has no concept of the relationship between the two.

In a relational database, we might make the relationship explicit in the schema using foreign key constraints, but Cassandra doesn't offer anything like this. In fact, if we want to use two different tables for `users` and `user_status_updates`, there isn't anything we can do to explicitly encode their relationship in the database schema. However, there is a way to combine user profiles and status updates into a single table, while still maintaining the one-to-many relationship between them. To achieve this merger, we'll use a feature of Cassandra tables that we haven't seen before — **static columns**.

Defining static columns

To see how static columns work, we'll create a new table called
`users_with_status_updates`. Our goal for this table is to contain both user
profiles and users' status updates; however, we only want one copy of a user
profile, even though each user might have many status updates. To accomplish
this, we'll add all of the columns from both `users` and `user_status_updates`
to our new table, but we'll declare the user profile columns as `STATIC`:

```
CREATE TABLE "users_with_status_updates" (
  "username" text,
  "id" timeuuid,
  "email" text STATIC,
  "encrypted_password" blob STATIC,
  "body" text,
  PRIMARY KEY ("username", "id")
);
```

Since the `email` and `encrypted_password` columns are properties of a user, not of a
specific status update, we declare them `STATIC`. Any column that is declared `STATIC`
has one value per partition key. In other words, there will be exactly one `email` value
associated with a given `username` in our table.

> The goal of static columns is to allow rows that share a partition key
> value to share other data as well. For this to be useful, there must be
> multiple rows per partition key, which is another way of saying that
> there must be at least one clustering column. It's illegal to declare a
> static column in a table with no clustering columns.

Working with static columns

Let's say that we've decided, in our MyStatus application, that when a user
creates a new account, we also require them to create an initial status update. In
this workflow, we'll end up creating a new user record at the same time as when
we create their first status update, which is to say that we'll be populating all the
columns of the `users_with_status_updates` table in a single statement:

```
INSERT INTO "users_with_status_updates"
("username", "id", "email", "encrypted_password", "body")
VALUES (
  'alice',
  76e7a4d0-e796-11e3-90ce-5f98e903bf02,
  'alice@gmail.com',
  0x8914977ed729792e403da53024c6069a9158b8c4,
  'Learning Cassandra!'
);
```

It's worth emphasizing that even though this is a single INSERT statement, we're actually doing two separate things:

- Assigning the email and encrypted_password values to every row whose username value is alice

- Creating a new row in the alice partition, with Learning Cassandra! as the value of body

Let's take a look at the contents of the table so far:

```
SELECT * FROM "users_with_status_updates";
```

Looking at the output, we'll observe that it's just what we'd expect (I've omitted the encrypted_password column to save space):

```
username | id                                   | email           | body
---------+--------------------------------------+-----------------+--------------------
   alice | 76e7a4d0-e796-11e3-90ce-5f98e903bf02 | alice@gmail.com | Learning Cassandra!

(1 rows)
```

The single row shown in the above figure is precisely what we wrote with the INSERT statement earlier. So far, we haven't seen any special behavior for static columns. To see static columns in action, let's add another status update to the table:

```
INSERT INTO "users_with_status_updates"
("username", "id", "body")
VALUES ('alice', NOW(), 'Another status update');
```

Note that in this case, we did not specify values for either of the static columns; we provided the partition key, the clustering column, and a value for the non-static data column body. Now, let's take a look at the contents of the table:

```
SELECT * FROM "users_with_status_updates";
```

The output is as follows:

```
username | id                                   | email           | body
---------+--------------------------------------+-----------------+----------------------
   alice | 76e7a4d0-e796-11e3-90ce-5f98e903bf02 | alice@gmail.com |   Learning Cassandra!
   alice | c5eeb290-7345-11e4-a0e6-5f98e903bf02 | alice@gmail.com | Another status update

(2 rows)
```

Now, the effect of the STATIC column declaration is clear: both rows have the email address we assigned to the first row even though we did not specify an email value in our second INSERT query.

Interacting only with the static columns

We've seen that static columns give us the ability to associate data with all rows that share the same partition key. However, what if we want to treat the data associated with that partition key as a discrete value rather than some extra information attached to each clustering column value?

More concretely, if we want to display an interface for alice to edit her user profile, all we really need to do is get the information associated with her partition key. We don't care if there are one, ten, or a thousand status updates; we just need a single row with her username, email address, and so on. With users as a separate table, this is trivial; we just select the single row with the partition key value alice. Let's see what happens when we do the equivalent in users_with_status_updates:

```
SELECT "username", "email", "encrypted_password"
FROM "users_with_status_updates"
WHERE "username" = 'alice';
```

Note that we've omitted the id and body fields, which contain distinct values in each row; we only select the partition key—username, and the static columns— email and encrypted_password. Unfortunately, we don't quite get what we're looking for:

```
username | email           | encrypted_password
---------+-----------------+------------------------------------------------------
   alice | alice@gmail.com | 0x8914977ed729792e403da53024c6069a9158b8c4
   alice | alice@gmail.com | 0x8914977ed729792e403da53024c6069a9158b8c4

(2 rows)
```

The result is two rows—the rows contain identical data because all of the selected columns are per-partition-key, but we still get duplicate results when we would only like one. Happily, the DISTINCT keyword allows us to retrieve a result in the format that we'd like:

```
SELECT DISTINCT "username", "email", "encrypted_password"
FROM "users_with_status_updates"
WHERE "username" = 'alice';
```

Now, the result takes the form we'd like—a single row containing the information specific to alice's partition key:

```
username | email             | encrypted_password
---------+-------------------+------------------------------------------
   alice | alice@gmail.com   | 0x8914977ed729792e403da53024c6069a9158b8c4

(1 rows)
```

By formulating our SELECT statement correctly, we can basically ignore the fact that there's a clustering column at all; the result here is exactly the same as it would have been if we had selected alice's row out of the users table. To complete this analogy, we'd also like to be able to write static columns to our table without thinking about the non-static columns.

Static-only inserts

As it turns out, inserting a row that consists only of static data is perfectly valid. Let's say, we relax our rule that upon signing up, a user must also write their first status update. If bob wishes to sign up, he may simply enter a username, email, and password:

```
INSERT INTO "users_with_status_updates"
("username", "email", "encrypted_password")
VALUES (
  'bob',
  'bob@gmail.com',
  0x10920941a69549d33aaee6116ed1f47e19b8e713
);
```

Now, we have an `email` and an `encrypted_password` associated with the partition key `bob`, but no clustering column values forming rows under that partition key. So, can we retrieve the information about `bob` that we just inserted? As it turns out, we can. Here's the updated state of the table that can be retrieved using the `SELECT *` `FROM "users_with_status_updates";` statement:

```
username | id                                   | email           | body
---------+--------------------------------------+-----------------+-----------------------
     bob |                                 null |   bob@gmail.com |                   null
   alice | 76e7a4d0-e796-11e3-90ce-5f98e903bf02 | alice@gmail.com |     Learning Cassandra!
   alice | c5eeb290-7345-11e4-a0e6-5f98e903bf02 | alice@gmail.com | Another status update
```

(3 rows)

We do, indeed, see `username` and `email` of `bob`, but we also notice something odd: the `id` column — part of the table's primary key — is `null`! This is because the row with the `username` value `bob` is essentially a synthetic row, representing the data associated with a partition key that has no actual rows. To satisfy ourselves of this fact, let's add a status update for `bob`:

```
INSERT INTO "users_with_status_updates"
("username", "id", "body")
VALUES ('bob', NOW(), 'Bob status update');
```

Now, let's take a look at the table again with the help of the `SELECT * FROM "users_with_status_updates";` statement:

```
username | id                                   | email           | body
---------+--------------------------------------+-----------------+-----------------------
     bob | 0b899e60-734a-11e4-a0e6-5f98e903bf02 |   bob@gmail.com |     Bob status update
   alice | 76e7a4d0-e796-11e3-90ce-5f98e903bf02 | alice@gmail.com |     Learning Cassandra!
   alice | c5eeb290-7345-11e4-a0e6-5f98e903bf02 | alice@gmail.com | Another status update
```

(3 rows)

The row with the `id` value `null` has disappeared, since the synthetic construct is no longer necessary. We now have access to static column values of `bob` via a real row.

 For more information about static columns, visit the page in the DataStax CQL reference at `http://www.datastax.com/documentation/cql/3.1/cql/cql_reference/refStaticCol.html`.

Static columns act like predefined joins

Now that we've explored the behavior of static columns, it's rather striking how similar the data model in `users_with_status_updates` is to the data model represented by `users` and `user_status_updates` taken together. In particular, both representations:

- Associate multiple status updates with a single user
- Associate a single email address and password with a single user
- Allow creating and accessing discrete user records, with or without associated status updates

The difference is simply the interface to the data model: are we working with one table, or two? It turns out that `users_with_status_updates` behaves very much like the result of an SQL LEFT JOIN of `users` and `user_status_updates`. Cassandra, of course, does not support JOIN clauses when reading data; instead, this join is baked into the schema itself through the use of static columns.

When to use static columns

We've now seen two ways to model users, each of whom has many status updates. The first approach was simply to define a table for users, and a table for their status updates; the relationship between the two is encoded in the key structure of `user_status_updates`. The second approach is to store all of the information in one `users_with_status_updates` table using static columns to associate user-level data with the `username` partition key rather than having a different value for each clustering column. Which is better?

The answer largely depends on how closely coupled the related data types are. If we expect that most of our interactions with user profile information will also involve interacting with the user's status updates, and vice versa, then it makes a lot of sense to store them together in one table. The reasoning is similar to the parent-child conceptual framework discussed earlier, but in reverse. Not only is the user a "special" relationship from the standpoint of the status update, but also the status update is a special relationship from the standpoint of the user.

In the case of MyStatus, on balance, the relationship between users and status updates doesn't quite pass this test. In subsequent chapters, we'll expand our schema to accommodate several different one-to-many relationships where the user plays the role of "parent". It's not clear that the relationship between users and status updates is more fundamental than the other relationships that we will model for users.

For this reason, we'll continue to work with the `users` and `user_status_updates` tables, and leave behind the `users_with_status_updates` table as an interesting exercise.

Refining our mental model

In the previous chapter, we began to develop a mental model of a Cassandra table that looked like a key-value store where each value is a collection of columns with values. Now that we have seen compound primary keys, we can refine that mental model to take into account the more complex structures we now know how to build, as follows:

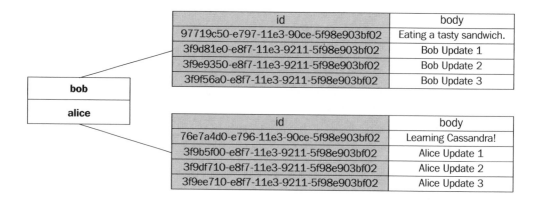

id	body
97719c50-e797-11e3-90ce-5f98e903bf02	Eating a tasty sandwich.
3f9d81e0-e8f7-11e3-9211-5f98e903bf02	Bob Update 1
3f9e9350-e8f7-11e3-9211-5f98e903bf02	Bob Update 2
3f9f56a0-e8f7-11e3-9211-5f98e903bf02	Bob Update 3

id	body
76e7a4d0-e796-11e3-90ce-5f98e903bf02	Learning Cassandra!
3f9b5f00-e8f7-11e3-9211-5f98e903bf02	Alice Update 1
3f9df710-e8f7-11e3-9211-5f98e903bf02	Alice Update 2
3f9ee710-e8f7-11e3-9211-5f98e903bf02	Alice Update 3

We can envision the `user_status_updates` table as a more robust key-value structure. Our keys are still usernames, but the values are now ordered collections of rows, both identified and ordered by the `id` clustering column. As with our earlier model, each partition key's data stands alone; to get data from multiple partitions, we have to go looking in multiple places.

Summary

In this chapter, we were introduced to the concept of compound primary keys, and learned that a primary key consists of one or more partition keys and, optionally, one or more clustering columns. We saw how partition keys—the only type of key we had previously encountered—can group related rows together, and how clustering columns provide an order for these rows within each partition.

Compound primary keys allow us to build a table containing users' status updates because they expose two important structures: grouping of related rows, and ordering of rows. In the `user_status_updates` table, we encoded the relationship between users and their status updates implicitly in the structure of the primary key; the partition key refers to the parent row in the `users` table. We also explored the use of static columns to make this relationship explicit, storing all the information about users and their status updates in a single table.

In *Chapter 4, Beyond Key-Value Lookup*, we will dive into new ways of interacting with data that is organized using compound primary keys, allowing us to build functionality in our application for viewing users' timelines in a variety of interesting ways.

4
Beyond Key-Value Lookup

In the last chapter, we explored a powerful data organization concept in Cassandra — the compound primary key. This allowed us to construct a table for status updates that satisfies two important requirements: related data is grouped together, and records have a built-in ordering within each group. So far, however, we only have two tools for querying data: either retrieve a single row using its full primary key, or retrieve all the rows in a table. In this chapter, we will dive into the more sophisticated queries that are available on tables with clustering columns, and use these queries to build the main reading interface for our MyStatus application. By the end of this chapter, you'll know:

- How to retrieve all rows within a single partition
- How to retrieve rows within a range of clustering column values
- How to paginate over rows in a single partition
- How to change the order of results
- How to store rows in descending order of clustering column
- How to paginate over all rows in a compound primary key table
- How to construct autocomplete functionality using a compound primary key table

Looking up rows by partition

The core reading experience of the MyStatus application will be an interface to read a given user's status updates. In order to do this, we need to be able to retrieve status updates for a given user from the user_status_updates table. As you might expect, this follows naturally from the CQL syntax we've seen in previous chapters:

```
SELECT * FROM "user_status_updates"
WHERE "username" = 'alice';
```

Previously, we've used the WHERE keyword to specify an exact value for a full primary key. In the preceding query, we only specify the partition key part of the primary key, which allows us to retrieve only those rows that we've asked for the partition:

```
username | id                                   | body
---------+--------------------------------------+--------------------
   alice | 76e7a4d0-e796-11e3-90ce-5f98e903bf02 | Learning Cassandra!
   alice | 3f9b5f00-e8f7-11e3-9211-5f98e903bf02 |     Alice Update 1
   alice | 3f9df710-e8f7-11e3-9211-5f98e903bf02 |     Alice Update 2
   alice | 3f9ee170-e8f7-11e3-9211-5f98e903bf02 |     Alice Update 3
```

In the results, we only see the rows whose username is alice. To emphasize what we discussed in the previous chapter, in the *Looking up a specific status update* section, this is a very efficient query. Cassandra stores all of alice's status updates together, already in order; so returning this view of the table is quite inexpensive.

The limits of the WHERE keyword

At this point, we've seen that you can look up rows by partition key alone, or by a combination of a partition key and a clustering column. We can easily imagine other ways to use WHERE, but it's not as flexible as we might hope.

Restricting by clustering column

In *Chapter 3, Organizing Related Data* you learned that any row in a table is uniquely identified by the combined values of its primary key columns. However, in the case of user_status_updates, the role of the username column is superfluous for the purposes of uniqueness; since id is a UUID, we know that it alone will uniquely identify the row on its own.

So, can we skip the username partition key and just look up rows by the id clustering column? Let's give it a shot:

```
SELECT * FROM "user_status_updates"
WHERE id = 3f9b5f00-e8f7-11e3-9211-5f98e903bf02;
```

This query is a syntactically valid CQL, and the WHERE clause does identify an existing row in the table—specifically, the status update whose body reads Alice Update 1.

This is not, however, a legal query. Instead, we see an error message as follows:

```
Bad Request: Cannot execute this query as it might involve data filtering
and thus may have unpredictable performance. If you want to execute this
query despite the performance unpredictability, use ALLOW FILTERING
```

Recalling our mental model for compound primary key tables, Cassandra organizes status updates like a key-value store, where the partition key acts as the lookup key. Without the partition key, Cassandra can't efficiently get to the row(s) you've specified. Instead, it would simply need to iterate internally over all the rows in the table, looking for rows that meet the conditions in the query. This sort of full table scan is extremely expensive for any table of non-trivial size, and Cassandra won't let you perform one unless you indicate with the ALLOW FILTERING directive appended to the end of the query that you know what you're getting yourself into.

Restricting by part of a partition key

Our status_update_replies table has a two-column partition key consisting of status_update_username and status_update_id. In some cases, we may want to, for instance, retrieve all of the replies to a certain user's status updates. It would be nice if we could query by part of the partition key.

Let's insert a row into status_update_replies, and then try to retrieve it using only the status_update_username column:

```
INSERT INTO "status_update_replies"
("status_update_username", "status_update_id", "id", "body")
VALUES(
  'alice',
  76e7a4d0-e796-11e3-90ce-5f98e903bf02,
  NOW(),
  'Good luck!'
);
```

Let's take a look at the table contents:

```
SELECT * FROM "status_update_replies"
WHERE "status_update_username" = 'alice';
```

As it turns out, this is also illegal:

```
Bad Request: Partition key part status_update_id must be restricted
since preceding part is
```

This error message tells us that if we're going to restrict any part of the partition key, we need to restrict all of it. This constraint is necessary because the layout of partition keys in a table is not related to their semantic ordering: instead, they're organized using the hash-like token of each key that we encountered in *Chapter 2, The First Table*. So, rows that contain `alice` in the `status_update_username` column will be distributed evenly through the partition key space; Cassandra doesn't have any way of knowing where to find them.

If we decide that retrieving all replies to a given user's status updates is an important access pattern, we can choose a different primary key arrangement for status update replies. If we made `status_update_username` alone the partition key, with `status_update_id` and `id` together forming the clustering column, then the resulting table would be ideally suited for the query in question. It would still be possible to look up the replies to a single status update, although it would not be quite as efficient as in our original formulation. This sort of reasoning, known as query-driven schema design, is critical to designing efficient Cassandra schemas; we will discuss it at length in the *Designing around queries* section of *Chapter 5, Establishing Relationships*.

Retrieving status updates for a specific time range

Having explored the limits of the WHERE keyword in CQL, let's return to the `user_status_updates` table. Suppose we'd like to build an "archive" feature for MyStatus that displays all of the user's status updates for a requested month. In CQL terms, we want to select a range of clustering columns; for instance, let's get back all of `alice`'s status updates created in May 2014:

```
SELECT "id", DATEOF("id"), "body"
FROM "user_status_updates"
WHERE "username" = 'alice'
AND "id" >= MINTIMEUUID('2014-05-01')
AND "id" <= MAXTIMEUUID('2014-05-31');
```

Before diving into the mechanics of this query, we can confirm that the only status update is the one with a UUID that was provided in the previous chapter, and was generated on May 29:

```
 id                                   | DATEOF(id)               | body
--------------------------------------+--------------------------+---------------------
 76e7a4d0-e796-11e3-90ce-5f98e903bf02 | 2014-05-29 21:05:17-0400 | Learning Cassandra!
```

Creating time UUID ranges

In the preceding query, we encounter two new CQL functions: MINTIMEUUID and MAXTIMEUUID. These functions form perhaps the most powerful components of Cassandra's toolkit for working with timestamp-based UUIDs.

As you learned in *Chapter 1*, *Getting Up and Running with Cassandra*, Version 1 UUIDs are generated using a timestamp, and this timestamp is the highest order consideration in how UUIDs are ordered. However, for each timestamp, there are about three hundred trillion possible UUIDs.

Since UUIDs are just numbers, it follows that for any given timestamp, there is a smallest UUID and a largest UUID. This is what the MINUUID and MAXUUID functions tell us. Cassandra doesn't give us a way to directly see the output of MINTIMEUUID and MAXTIMEUUID, but we can use them in the WHERE clauses as shown in the preceding query.

> For a complete account of the UUID-related functions available in CQL, refer to the DataStax CQL reference: http://www.datastax.com/documentation/cql/3.1/cql/cql_reference/timeuuid_functions_r.html

Selecting a slice of a partition

With a solid understanding of the UUID functions we're using in this query, we can now break down exactly what we're asking Cassandra to give us:

- Rows in the alice partition
- With an id that is greater than or equal to the minimum possible UUID generated on May 1, 2014
- With an id that is less than or equal to the maximum possible UUID generated on May 31, 2014

This is a very common query pattern in Cassandra, and is referred to as a **range slice query**. A range slice query selects a single partition key, and a range of clustering column values. Like a query that is purely restricted by partition key, range slice queries are very efficient as they take advantage of the underlying layout of the data in the table.

Paginating over rows in a partition

As our users create more and more status updates, we'll build pagination functionality into MyStatus so that the information on the page doesn't overwhelm readers. For the sake of convenience, let's say that each page will only contain three status updates.

To retrieve the first page, we'll use the LIMIT keyword that we first encountered in *Chapter 2*, *The First Table*:

```
SELECT "id", DATEOF("id"), "body"
FROM "user_status_updates"
WHERE "username" = 'alice'
LIMIT 3;
```

As expected, Cassandra will give us the first three rows in ascending order of id:

```
 id                                   | DATEOF(id)                | body
--------------------------------------+---------------------------+---------------------
 76e7a4d0-e796-11e3-90ce-5f98e903bf02 | 2014-05-29 21:05:17-0400 | Learning Cassandra!
 3f9b5f00-e8f7-11e3-9211-5f98e903bf02 | 2014-05-31 15:10:37-0400 |      Alice Update 1
 3f9df710-e8f7-11e3-9211-5f98e903bf02 | 2014-05-31 15:10:37-0400 |      Alice Update 2
```

Now, we'll ask for the collection of rows where the id value is strictly greater than the last id we saw:

```
SELECT "id", DATEOF("id"), "body"
FROM "user_status_updates"
WHERE "username" = 'alice'
  AND id > 3f9df710-e8f7-11e3-9211-5f98e903bf02
LIMIT 3;
```

This is similar to the pagination query for users that we made in *Chapter 2*, *The First Table*, but it's in some ways simpler. We keep the restriction of rows to alice's partition, and then add a simple greater than restriction for id.

Unlike in the users table, we don't need to use the TOKEN function, because id is not a partition key. Since clustering columns are stored in semantic order rather than token order, we just require the next set of logically greater id values. The result is the fourth and final row in alice's status updates:

```
 id                                   | DATEOF(id)                | body
--------------------------------------+---------------------------+-----------------
 3f9ee170-e8f7-11e3-9211-5f98e903bf02 | 2014-05-31 15:10:37-0400 | Alice Update 3
```

Counting rows

While in some cases, it might be fine to simply keep requesting pages until no more records are encountered, it's often nice to give the user a sense of how many records they're dealing with up front. Happily, we can use the COUNT function to do just this:

```
SELECT COUNT(1) FROM "user_status_updates"
WHERE "username" = 'alice';
```

Instead of requesting a list of columns or a wildcard to get all columns, we select COUNT(1) to get a total count of rows matching the WHERE condition:

```
 count
-------
     4

(1 rows)
```

We could have also used COUNT(*), which behaves in the same way as COUNT(1). There are, however, no other valid ways to use the COUNT function in CQL.

> While the COUNT function is convenient, it's of dubious utility in production applications. Under the hood, performing a COUNT query requires the same work for Cassandra that actually returning the rows would; by using COUNT, you're mostly saving network bandwidth. If you need to efficiently track the number of rows stored in a given table or partition, you're better off using a counter cache. We'll introduce a good way to do this in *Chapter 9, Aggregating Time-Series Data*. For more information on COUNT queries, refer to the DataStax Cassandra documentation for SELECT: http://www.datastax.com/documentation/cql/3.1/cql/cql_reference/select_r.html?scroll=reference_ds_d35_v2q_xj__counting-returned-rows

Reversing the order of rows

We're on our way to building great data access logic for displaying a user's timeline of status updates. There's one minor problem though; if I'm viewing alice's status updates, I'm probably interested in reading her most recent ones first. So far, we've always gotten the status updates in ascending order of id, meaning the oldest ones come first. Fortunately, we've got a couple of ways to change this.

Reversing clustering order at query time

Using our existing `user_status_updates` table, we can instruct Cassandra to return results in reverse order of `id`:

```
SELECT "id", DATEOF("id"), "body"
FROM "user_status_updates"
WHERE "username" = 'alice'
ORDER BY "id" DESC;
```

This is the first time we've seen an ORDER BY in CQL, but it should be familiar to anyone who's worked with a SQL database: the DESC tells Cassandra that we want to order rows by descending values in the `id` column:

```
id                                   | DATEOF(id)               | body
-------------------------------------+--------------------------+---------------------
3f9ee170-e8f7-11e3-9211-5f98e903bf02 | 2014-05-31 15:10:37-0400 |       Alice Update 3
3f9df710-e8f7-11e3-9211-5f98e903bf02 | 2014-05-31 15:10:37-0400 |       Alice Update 2
3f9b5f00-e8f7-11e3-9211-5f98e903bf02 | 2014-05-31 15:10:37-0400 |       Alice Update 1
76e7a4d0-e796-11e3-90ce-5f98e903bf02 | 2014-05-29 21:05:17-0400 | Learning Cassandra!
```

You might assume that the ORDER BY gives us a lot of flexibility in ordering rows—you might think that `id` can just as well be replaced by any other column. Let's see for ourselves:

```
SELECT "id", DATEOF("id"), "body"
FROM "user_status_updates"
WHERE "username" = 'alice'
ORDER BY "body" DESC;
```

Here, we're trying to sort our rows by descending (Z-A) values in the `body` field. Alas, it is not to be:

```
Bad Request: Order by is currently only supported on the clustered
columns of the PRIMARY KEY, got body
```

As it turns out, the *only* valid argument to ORDER BY clause is the name of the first clustering column. Effectively, the entire ORDER BY clause exists purely to allow you to reverse the order of rows from their natural clustering order.

Reversing clustering order in the schema

While it's useful to be able to show the newest status updates first using ORDER BY, it would be even better if the rows were naturally ordered newest first. Although we can't change the clustering order of an existing table, we can make a new table that has the same structure as user_status_updates, but with id ordered newest first:

```
CREATE TABLE "reversed_user_status_updates" (
  "username" text,
  "id" timeuuid,
  "body" text,
  PRIMARY KEY ("username", "id")
) WITH CLUSTERING ORDER BY ("id" DESC);
```

The CLUSTERING ORDER BY property tells Cassandra that we want to store columns in descending order by id, rather than the default ascending order. Adding a few quick rows to our table, we can see the reversed clustering order in action:

```
INSERT INTO "reversed_user_status_updates"
("username", "id", "body")
VALUES ('alice', NOW(), 'Reversed status 1');
INSERT INTO "reversed_user_status_updates"
("username", "id", "body")
VALUES ('alice', NOW(), 'Reversed status 2');
INSERT INTO "reversed_user_status_updates"
("username", "id", "body")
VALUES ('alice', NOW(), 'Reversed status 3');
SELECT * FROM "reversed_user_status_updates"
WHERE "username" = 'alice';
```

Even though we did not specify a descending order, the rows are naturally returned with the newest records first:

```
username | id                                   | body
---------+--------------------------------------+------------------
   alice | 64e9cc20-f03a-11e3-995c-5f98e903bf02 | Reversed status 3
   alice | 64e956f0-f03a-11e3-995c-5f98e903bf02 | Reversed status 2
   alice | 64e670c0-f03a-11e3-995c-5f98e903bf02 | Reversed status 1
```

There are a couple of reasons for which you might want to specify an explicit clustering order at table creation time. First, reversing the clustering order at query time has a slight performance penalty—you don't need to avoid it but all things being equal, it's best to create the table with the clustering order corresponding to your most frequent access pattern. In the case of status updates, our second table with the reversed clustering order is better, since in most cases we'll want to see the newest status updates first.

Second, if you have multiple clustering columns, you might want to order by a mixture of ascending and descending columns. For instance, you might want to store a list of products in ascending order by manufacturer name, and then in a descending order by the date on which the product was added within each manufacturer. In this case, you would need to set the CLUSTERING ORDER at table creation time, since ORDER BY does not allow you to reverse the clustering order of one column without reversing the order of all columns.

Paginating over multiple partitions

Returning to our original user_status_updates table, we might in certain situations, for an administrative interface, for instance, display all of the status updates in the system. In this case, we will certainly want the ability to paginate, as the total collection of status updates will get very large.

As in our previous example, let's use a page size of three. The query for the first page is simple enough:

```
SELECT * FROM "user_status_updates"
LIMIT 3;
```

As expected, we'll get back bob's first three status updates:

```
 username | id                                   | body
----------+--------------------------------------+-------------------------
      bob | 97719c50-e797-11e3-90ce-5f98e903bf02 | Eating a tasty sandwich.
      bob | 3f9d81e0-e8f7-11e3-9211-5f98e903bf02 |            Bob Update 1
      bob | 3f9e9350-e8f7-11e3-9211-5f98e903bf02 |            Bob Update 2
```

For the second page, things get a bit more complicated. We know that the last row we retrieved was from bob's partition, but we're not sure if there are any more status updates for bob. In case there are, we will ask for the next page of bob's partition:

```
SELECT * FROM "user_status_updates"
WHERE "username" = 'bob'
  AND id > 3f9e9350-e8f7-11e3-9211-5f98e903bf02;
```

In this query, we ask for any status updates in `bob`'s partition that have an `id` that is greater than the last `id` we saw in the previous page. As it turns out, there is one more row:

```
username | id                                   | body
---------+--------------------------------------+-------------
     bob | 3f9f56a0-e8f7-11e3-9211-5f98e903bf02 | Bob Update 3
```

Since we want to display pages of three, we should try to fetch more results before displaying this page to the user. Knowing that we've reached the end of `bob`'s partition, let's see if we can find two more results in partitions that are situated after `bob`'s:

```
SELECT * FROM "user_status_updates"
WHERE TOKEN("username") > TOKEN('bob')
LIMIT 2;
```

By asking for usernames whose partition token is greater than `bob`'s token, we guarantee that we get back results we haven't seen yet. The rows returned are the first two in `alice`'s status updates:

```
username | id                                   | body
---------+--------------------------------------+---------------------
   alice | 76e7a4d0-e796-11e3-90ce-5f98e903bf02 | Learning Cassandra!
   alice | 3f9b5f00-e8f7-11e3-9211-5f98e903bf02 |       Alice Update 1
```

Since we requested two rows and retrieved two rows, we know that there might be more rows in `alice`'s partition. Once again, we ask for `id` values greater than the last one that we saw in `alice`'s status updates:

```
SELECT * FROM "user_status_updates"
WHERE "username" = 'alice'
   AND id > 3f9b5f00-e8f7-11e3-9211-5f98e903bf02
LIMIT 3;
```

The query returns two remaining status updates for `alice`:

```
username | id                                   | body
---------+--------------------------------------+----------------
   alice | 3f9df710-e8f7-11e3-9211-5f98e903bf02 | Alice Update 2
   alice | 3f9ee170-e8f7-11e3-9211-5f98e903bf02 | Alice Update 3
```

Since we asked for three rows and only got back two, we know that we've now seen all the status updates in `alice`'s partition. Now, we should check if there are any more partitions to iterate over; we'll just ask for one, since we already have the first two rows of this page:

```
SELECT * FROM "user_status_updates"
WHERE TOKEN("username") > TOKEN('alice')
LIMIT 1;
```

When we run this query, we'll find that no rows are returned. This tells us that we've reached the last page.

This sort of two-step pagination might seem a bit arduous, but the process can be expressed concisely. We followed these steps until no more rows were returned:

1. Ask for the first rows in the table.

2. Ask for the next rows in the same partition as the last row that was returned. Repeat until no more rows are returned for this partition.

3. Ask for rows in the next partition.

4. Go back to step 2.

Building an autocomplete function

So far, we've been focused on storing users and their status updates, but we can use our knowledge of compound primary keys to make it a bit easier to write status updates too. Let's introduce a hashtagging function into the status update composition interface, and then autocomplete hashtags as users type them.

First, we'll set up a table to store hashtags using the following query:

```
CREATE TABLE "hash_tags" (
  "prefix" text,
  "remaining" text,
  "tag" text,
  PRIMARY KEY ("prefix", "remaining")
);
```

The structure of our table is a bit unusual but it will work very well for our purposes. The partition key is `prefix`, which we'll use to store the first two letters of each hashtag. The clustering column, `remaining`, will store the remaining letters of the hashtag, and `tag` will contain the entire hashtag start to finish.

By partitioning the table this way, we'll make things easy for Cassandra by immediately narrowing down the list of possible autocomplete tags to those in the partition, identified by the two-letter prefix. This does, of course, mean that we can't autocomplete entries of fewer than two letters, but this is a pretty standard limitation.

To see how the autocomplete works, let's add some seed data into the table:

```
INSERT INTO "hash_tags" ("prefix", "remaining", "tag")
VALUES ('ca', 'ssandra', 'cassandra');
INSERT INTO "hash_tags" ("prefix", "remaining", "tag")
VALUES ('ca', 'ssette', 'cassette');
INSERT INTO "hash_tags" ("prefix", "remaining", "tag")
VALUES ('ca', 'sual', 'casual');
INSERT INTO "hash_tags" ("prefix", "remaining", "tag")
VALUES ('ca', 'ke', 'cake');
```

Now, let's say the user starts typing a hashtag, and has so far typed ca. That's easy: we can just ask for the ca partition in our table:

```
SELECT "tag" FROM "hash_tags"
WHERE "prefix" = 'ca';
```

Just as we want, we'll get tags starting with the letters ca:

```
      tag
----------
     cake
cassandra
 cassette
   casual
```

So far so good! Now, let's say the user has typed cas. In this case, we want to return all the words starting with cas, but CQL doesn't have a starting-with operator. Instead, we'll use a range slice query to describe the rows we're looking for. From the standpoint of lexical string ordering, we're looking for strings that are lexically greater than cas, but lexically smaller than cat, since t is the next letter after s. Translated into a CQL query, it looks like this:

```
SELECT "tag" FROM "hash_tags"
WHERE "prefix" = 'ca'
  AND "remaining" >= 's'
  AND "remaining" < 't';
```

The remaining range slice will cover strings like `ssete`, `sual`, and `ssandra`, because those are all lexically greater than `s` and lexically smaller than `t`. So, we get just the results we're looking for:

```
          tag
       ----------
       cassandra
        cassette
         casual
```

Supposing the user typed a couple more letters, for a total entry of `cassa`, we'd narrow the range further:

```
SELECT "tag" FROM "hash_tags"
WHERE "prefix" = 'ca'
  AND "remaining" >= 'ssa'
  AND "remaining" < 'ssb';
```

Again, we construct the upper bound for the range slice by simply replacing the last character of the input with the following character in the alphabet, giving us the right lexical upper bound. Now we've narrowed our results to a single hashtag:

```
          tag
       ----------
       cassandra
```

While compound primary keys are an intuitively good fit storing timestamped data partitioned into logical groups, with a little creativity the data structure can be applied to a wide range of other problems. The autocomplete implementation we built here might not be the simplest one you've run across, but it can efficiently store and autocomplete billions of hashtags, which isn't anything to sneeze at!

Summary

In this chapter, we've explored data access patterns for compound primary key tables, and used our new knowledge to expose a paginated view of a user's most recent status updates. We discussed how to query range slices of clustering column values, and how to reverse the clustering order at both table creation time and at query time. We reinforced our understanding of the CQL SELECT queries by performing the intricate task of paginating over all the rows in our status updates table, and then applied our new compound primary key toolkit to a totally different problem, the autocompletion of hashtags.

In *Chapter 5, Establishing Relationships*, we will apply our techniques for compound primary key data modeling to the task of describing relationships that do not have the clear parent-child structure of users and status updates. We'll also go beyond primary key-based data access, exploring the use of secondary indexes to look up data using any column we wish.

5

Establishing Relationships

At this point, we might declare our MyStatus application a minimum viable product. Users can create accounts and post status updates, and those status updates can be viewed in their authors' timelines. And, of course, since we're storing the data in Cassandra, we don't need to worry about scaling up to millions of users or billions of status updates.

As our service grows, however, it would be nice for users to be able to view all their friends' status updates in one place. The first step of that, of course, would be to know who a user's friends are. So, we'll build a feature that allows one user to follow another.

We've already seen a good way to use Cassandra to model a specific type of relationship. Compound primary keys are a natural fit for parent-child associations but a follow relationship is many-to-many: I follow many users, and—hopefully—many users follow me.

In this chapter, we'll build on our knowledge of Cassandra data modeling and introduce new patterns to model relationships. We'll explore two ways to model relationships. First, we will explicitly create a table whose role is to represent the relationship between two other entities. Second, we'll look at how secondary indexes can help us represent relationships without the need for an extra lookup table.

By the end, we'll have two alternative data structures to capture followers in the database, and you'll have learned:

- How to model many-to-many relationships in Cassandra
- How query-driven design can produce denormalized data structures
- How to delete rows from a Cassandra table
- How to create secondary indexes
- The limitations and disadvantages of secondary indexes

Modeling follow relationships

A data model for **follow** relationships should be able to answer two questions on behalf of a user:

- Who do I follow?
- Who follows me?

In *Chapter 3, Organizing Related Data,* you learned to design our table structures so that all important data access can be accomplished by querying a single partition. For this reason, we're better off considering the above questions separately, and designing the right table schema for each.

Outbound follows

We'll start with the question, "Who do I follow?" We'll want a partition per user, with each partition containing all the other users they follow:

```
CREATE TABLE "user_outbound_follows" (
  "follower_username" text,
  "followed_username" text,
  PRIMARY KEY ("follower_username", "followed_username")
);
```

Simple enough, but there's something unusual here: there are only two columns in the table, and they're both part of the primary key. As it turns out, this is a perfectly valid way to construct a table schema; non-key columns are optional in Cassandra tables. Since all we need to know is who's at the other end of the follow relationship, which is available via the clustering column `followed_user_id`, no additional data columns are needed.

Inbound follows

Our new `user_outbound_follows` table allows us to efficiently find out who a user follows; since the partition key is the username of the following user, we'll only be accessing a single partition to answer the question. It does not, however, provide any way to find out who follows a certain user. In order to broadcast users' status updates to their followers, we'll certainly need to be able to find out who their followers are.

To accomplish this, we'll create a new table, which is the mirror image of `user_outbound_follows`:

```
CREATE TABLE "user_inbound_follows" (

    "followed_username" text,
    "follower_username" text,
    PRIMARY KEY ("followed_username", "follower_username")
);
```

The new `user_inbound_follows` table looks exactly like the `user_outbound_follows` table, except that the order of the keys is reversed: now the followed user's username is the partition key, and each row within the partition is identified by the follower user's username. Like `user_outbound_follows`, our new table does not have any non-key columns.

Storing follow relationships

We've now created two tables, each of which allows us to answer an important question about follow relationships: first, whom does a user follow; and second, who follows a user. Now let's establish some follow relationships.

For now, let's have `alice` follow a couple of other users, `bob` and `carol`:

```
INSERT INTO "user_outbound_follows"
    ("follower_username", "followed_username")
VALUES ('alice', 'bob');

INSERT INTO "user_inbound_follows"
    ("followed_username", "follower_username")
VALUES ('bob', 'alice');

INSERT INTO "user_outbound_follows"
    ("follower_username", "followed_username")
VALUES ('alice', 'carol');

INSERT INTO "user_inbound_follows"
    ("followed_username", "follower_username")
VALUES ('carol', 'alice');
```

For each follow relationship, we have to insert two rows: one in the `user_outbound_follows` table to store the relationship from the perspective of the follower, and one in the `user_inbound_follows` table to store the relationship from the perspective of the followed user.

Designing around queries

Our choice to store two representations of a single follow relationship, is a prime example of **query-driven design**: the practice of designing table schemas to accommodate the data access patterns of our application, rather than simply building structures to naturally represent the underlying data. In the case of follow relationships, either the `user_outbound_follows` or `user_inbound_follows` table is sufficient on its own to store the data about follow relationships; were we using a relational database, it is likely that we would only use one of these tables, with secondary indexes on the non-key columns for efficient lookup. But using Cassandra, neither table is able to satisfy all of the query needs our application will have, so query-driven design instructs us to include both in our database.

When thinking about the best way to satisfy the data access needs of our application, it's helpful to think of our Cassandra tables in terms of core data structure: a map of values to collections of values. If we want to know who a given user follows, we want a map of usernames to the collection of users they follow. This abstract representation of the ideal data structure instructs us to build a table whose primary key is a username, and whose rows consist of usernames of users they follow.

Denormalization

Our follow tables are also the first example we've seen of **denormalization**, which is the practice of storing the same data in more than one place. Denormalization is typically frowned upon in relational database schemas, although from a practical standpoint it's often a useful optimization even in that scenario. In non-relational databases, denormalization is often a critical tool in query-driven design.

The downside of denormalization is exemplified by our preceding insert pattern: we have to make two INSERT statements to fully represent one fundamental fact. From a standpoint of performance, this is acceptable: Cassandra is optimized for efficient write operations, so we're happy to make verbose writes in order to allow efficient reads. This does, of course, add more complexity at the application level: the application is responsible to ensure that any modification to the `user_outbound_follows` table is accompanied by an equivalent modification to the `user_inbound_follows` table. If there is a flaw in the application's logic, the two tables may not contain consistent information.

Looking up follow relationships

Now that we've studiously designed our follow tables to efficiently support our application's data access patterns, let's do some data access. To start, we'll want to give `alice` an interface to manage the list of users she follows; this interface will, of course, need to show her who she currently follows:

```
SELECT "followed_username"
FROM "user_outbound_follows"
WHERE "follower_username" = 'alice';
```

Here we ask for all of the outbound follows in the partition of `alice`: an efficient query, since it only looks up a single partition's worth of data. As expected, we see that `alice` follows `bob` and `carol`:

```
         followed_username
        --------------------
                        bob
                      carol


        (2 rows)
```

Note that the usernames returned are in alphabetical order: this is not a coincidence. Since `followed_username` is the clustering column in the `user_outbound_follows` table, the rows are stored in string order of the followed user's username. While this isn't critical to the functionality of our application, it's a happy *bonus* feature of the data structure we've chosen.

Now let's explore the other side of the relationship, "Who follows `bob`?" To reinforce our earlier discussion of query-driven design, let's first try to answer that question using the `user_outbound_follows` table:

```
SELECT "follower_username"
FROM "user_outbound_follows"
WHERE "followed_username" = 'bob';
```

When we attempt this query, we see an error that is familiar from the previous chapter: Cassandra is unable to efficiently perform the lookup we're asking for, so we're required to include the ALLOW FILTERING directive as a *waiver* indicating that we know the query may be expensive:

```
Bad Request: Cannot execute this query as it might involve data filtering
    and thus may have unpredictable performance. If you want to execute this
    query despite the performance unpredictability, use ALLOW FILTERING
```

Having assured ourselves that our denormalized structure is justified, we'll use the `user_inbound_follows` table to find out who follows `bob`:

```
SELECT "follower_username"
FROM "user_inbound_follows"
WHERE "followed_username" = 'bob';
```

Since we're doing the lookup using the `follower_username` partition key, Cassandra has no complaints, and the result is what we expect:

```
     follower_username
    --------------------
                  alice

(1 rows)
```

Poor `bob` has only one follower. Keep posting those funny cat pictures, `bob`, and you'll have more in no time!

Unfollowing users

It's conceivable that `bob` may end up posting too many funny cat pictures for `alice`'s taste, in which case, she may decide to unfollow him. For `alice` to do that, we'll need to remove the rows representing the follow relationship from both the inbound and outbound follow tables:

```
DELETE FROM "user_outbound_follows"
WHERE "follower_username" = 'alice'
  AND "followed_username" = 'bob';
```

```
DELETE FROM "user_inbound_follows"
WHERE "followed_username" = 'bob'
  AND "follower_username" = 'alice';
```

This is our first encounter with CQL's DELETE statement, although it should look quite familiar to anyone who's worked with SQL. To delete a row, we specify the full primary key of the row, which is to say both the partition key(s) and the clustering column(s). The WHERE...AND syntax is the same as that used in SELECT queries, introduced in *Chapter 3, Organizing Related Data*.

To check the effects of the deletion, we can query again for the list of users `alice` follows:

```
SELECT "followed_username"
FROM "user_outbound_follows"
WHERE "follower_username" = 'alice';
```

As expected, `alice` no longer follows `bob`:

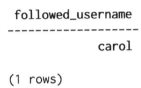

```
    followed_username
   -------------------
           carol

   (1 rows)
```

Were we to check the `user_inbound_follows` table for `bob`, we would find that he no longer has any followers.

 For more on the DELETE statement, see the DataStax CQL reference: http://www.datastax.com/documentation/cql/3.1/cql/cql_reference/delete_r.html

Using secondary indexes to avoid denormalization

So far, we've exclusively used primary key columns to look up rows—either the full primary key when we're looking for a specific row, or just the partition key when retrieving multiple rows in a single partition. We know that these kinds of lookups are very efficient, because Cassandra can satisfy the query by accessing the single region of storage that holds the partition's data in order.

This is the motivation for the denormalized follow structure we've built in this chapter: whether we want to answer the question, "Who does `alice` follow?", or the question, "Who follows `alice`?", we can construct a query that only needs to access a single partition. However, we're accepting additional complexity in the form of storing two versions of the same information, in `user_inbound_follows` and `user_outbound_follows`.

As it happens, Cassandra does provide us with a way to answer both questions in a reasonably efficient way using a single table, with a single representation of each follow relationship. By adding a **secondary index**, we can enable lookup of rows using columns other than the primary key.

The form of the single table

To demonstrate this approach, we'll start by creating a new table, `user_follows`, that has an identical structure to our `user_inbound_follows` table:

```
CREATE TABLE "user_follows" (
  "followed_username" text,
  "follower_username" text,
  PRIMARY KEY ("followed_username", "follower_username")
);
```

As with `user_outbound_follows`, this new table uses a followed user as its partition key, and the follower as its clustering column. At a higher level, for each user, *U*, this table stores a partition of rows representing the users that follow *U*. So, if we'd like to find out who follows a given user, we can do so very efficiently using this table.

However, we've said that we want to use the same table for both in bound and out bound follows; that is to say, we'd like this table to be able to answer the question, "Who does Alice follow?" Let's recall what happens if we naively try a query to answer that question on the `user_follows` table:

```
SELECT * FROM "user_follows"
WHERE "follower_username" = 'alice';
```

Of course, Cassandra won't allow us to perform this query:

```
code=2200 [Invalid query] message="Cannot execute this query as it might
 involve data filtering and thus may have unpredictable performance. If
you want to execute this query despite the performance unpredictability,
 use ALLOW FILTERING"
```

The structure of Cassandra's primary keys requires us to strictly adhere to a set of rules for what columns we may specify in the WHERE clause of a query. In particular, if we want to look up rows by a given column or columns, we can use:

- Exact values for all partition key columns
- Exact values for all partition key columns, and exact value for the first clustering column
- Exact values for all partition key columns, and exact values for the first two clustering columns

Most importantly, if we're going to specify values for any columns, we must specify values for the partition key columns. This is because the partition key is the top-level lookup key for rows in any table; recall our mental model from the *Refining our mental model* section at the end of *Chapter 3, Organizing Related Data.*

Adding a secondary index

There is one exception to this rule, and that is when a secondary index is in use. A secondary index should be a familiar concept to anyone who has worked with a relational database; in the relational world, they're often just referred to as indexes. Put simply, secondary indexes allow us to perform a reasonably efficient lookup of rows using columns other than the partition key.

In our case, we'd like to look up rows in `user_follows` by providing a value for the `follower_username` row but omitting a value for the `followed_username` partition key. To do this, we can use the `CREATE INDEX` statement:

```
CREATE INDEX ON "user_follows" ("follower_username");
```

The syntax here is pretty straightforward: we simply provide Cassandra with the name of a table, and the name of a column within the table that we'd like to create an index on.

To see our index in action, let's add a few follow relationships to our new table:

```
INSERT INTO "user_follows"
  ("followed_username", "follower_username")
VALUES ('alice', 'bob');


INSERT INTO "user_follows"
  ("followed_username", "follower_username")
VALUES ('alice', 'carol');


INSERT INTO "user_follows"
  ("followed_username", "follower_username")
VALUES ('carol', 'bob');


INSERT INTO "user_follows"
  ("followed_username", "follower_username")
VALUES ('dave', 'bob');
```

Now let's say we'd like to answer the question, "Who does bob follow?" We can construct a query of the same form that caused us an error before we created the index:

```
SELECT * FROM "user_follows"
WHERE "follower_username" = 'bob';
```

But this time, we'll get the results we're looking for:

```
followed_username | follower_username
------------------+------------------
           dave   |              bob
           carol  |              bob
           alice  |              bob

(3 rows)
```

This is the first time we've been able to construct a query that looks up rows by a column value without specifying the partition key. Secondary indexes give a significant new tool when designing our data models and access patterns.

 In this case, we created a secondary index on a clustering column, but there's no restriction on the kind of column that can be used for a secondary index. We'll explore more examples here and in *Chapter 8, Collections, Tuples, and User-defined Types*.

Other uses of secondary indexes

We've seen that secondary indexes neatly allow us to turn a denormalized relationship structure into a normalized one: the user_follows table is able to answer all the important questions about follow relationships without any duplication of data. This is but one of the many use cases for which secondary indexes are well suited.

Secondary indexes are best suited for **low-cardinality** columns, which is to say columns that contain the same value for many rows. An example might be a location column on the users table; if this is restricted to city and state, many users will share the same location. In fact, we will add a location column to the users table in *Adding columns to tables* section in *Chapter 7, Expanding Your Data Model*. If we wanted to be able to answer questions such as "Who are all of the users that live in New York?" that index would be quite useful.

Secondary indexes can also be used for columns whose values are unique, such as the email column in the users table. If, for instance, we wanted to build a "forgot password" feature in which the user enters their email address, we'd be able to use an index on email to look up the user's record.

Use caution when creating indexes on unique or high-cardinality columns; these indexes can only handle moderate query volume. The DataStax Cassandra reference warns us:

> *"For columns containing unique data, it is sometimes fine performance-wise to use an index for convenience, as long as the query volume to the table having an indexed column is moderate and not under constant load."*

Another interesting use case for secondary indexes would be for easier lookup of status updates from the `user_status_updates` table. Recall that that table's partition key is the username of the user who wrote the status update, and the clustering column is a UUID column. In order to look up a specific status update, we would normally have to specify both `username` and `id` values. But the `id` column alone uniquely identifies each status update, because UUIDs are globally unique. So, if we place a secondary index on the `id` column, we can look up status updates more tersely, which is useful to generate URLs and the like.

> For more information on secondary indexes, see the DataStax CQL reference for the `CREATE INDEX` statement: `http://www.datastax.com/documentation/cql/3.1/cql/cql_reference/create_index_r.html`

Limitations of secondary indexes

We've seen that secondary indexes are useful in a range of scenarios, and we'll explore more in *Chapter 8, Collections, Tuples, and User-defined Types*. However, indexes are not without limitations and downsides.

Secondary indexes can only have one column

One major limitation of secondary indexes is that they can only target one column. It would not be legal, for instance, to specify a secondary index such as:

```
CREATE INDEX ON "users" ("email", "encrypted_password");
```

We'll explore one workaround to this limitation in the *Working with tuples* section of *Chapter 8, Collections, Tuples, and User-defined Types*.

Secondary indexes can only be tested for equality

In the section on *Retrieving status updates for a specific time range* in *Chapter 4, Beyond Key-Value Lookup*, we explored the use of inequality operators to select ranges of columns. Secondary indexes can only be queried for equality; queries such as the following are not possible:

```
SELECT * FROM "user_follows"
WHERE "follower_username" > 'alice';
```

Secondary index lookup is not as efficient as primary key lookup

Although secondary indexes give us a reasonably efficient way to look up rows using a non-partition key column, they're not as efficient as queries based on the primary key we've explored in previous chapters. This is because lookup by a secondary index is a two-step process. First, Cassandra will access the secondary index to find the primary keys of all rows matching the query. Second, it will access the table itself to retrieve the matched rows. The second step will generally involve querying over many partitions, or at least over disjoint ranges of a single partition. These random reads over the data can never be as efficient as a single focused read of one range of one partition.

For this reason, it's best to avoid using secondary indexes for the core data access patterns of your application. Our structure in the `user_follows` table bears this out: answering the question, "Who follows `alice`?", is central to the process of broadcasting her status updates to her followers. On the other hand, the question, "Who does `alice` follow?" is only necessary when `alice` or another user wishes to see a list of those users. By structuring the table with the `followed_username` column as the partition key, we ensure that super-efficient partition key lookup is available for the most important question, and use secondary index lookup for less essential queries.

> For more on when to use secondary indexes, and when not to, see the chapter from the DataStax Cassandra documentation: `http://www.datastax.com/documentation/cql/3.1/cql/ddl/ddl_when_use_index_c.html#concept_ds_sgh_yzz_zj__when-no-index`

Summary

In this chapter, we looked at ways to model relationships between objects that go beyond the straightforward parent-child relationships that are captured elegantly by a compound primary key. We found that query-driven schema design motivated us to create multiple representations of the follow relationship; each representation optimized to answer a specific question about follows. This led us to a denormalized schema, wherein each follow has multiple representations in our database.

While our denormalized schema requires more write operations than a normalized one, and extra care at the application level to ensure the different representations of follows are consistent with one another, we end up with better overall performance because writing data to Cassandra is cheaper than reading it. By designing our schema to allow Cassandra to efficiently access data in a single partition to answer any question the application needs, we ensure that Cassandra can continue efficiently serving the application's needs even at a massive scale.

As an alternative to denormalization, we introduced secondary indexes as a way to avoid storing the same follow relationship in two places. We explored the uses and limitations of secondary indexes, and recognized that, while it is convenient to only have a single table for follow relationships, we do lose some performance when performing a query to find out who a user follows.

Finally, we introduced deletion of rows, our first foray into modifying data that we've stored in Cassandra.

In the next chapter, we'll continue our explanation of denormalized data structures when we create home timelines for each user, which will allow us to broadcast a user's status updates to all of their followers.

6
Denormalizing Data for Maximum Performance

In the previous chapter, we created a structure that allows a user to follow other users. The goal of the follow system was to allow users to see all of their followed users' status updates in one place that we'll call the "home timeline". In this chapter, we will build a table to store users' home timelines.

The follow structures in *Chapter 5, Establishing Relationships* introduced the concept of denormalization, the practice of storing the same piece of data in more than one place in order to optimize read performance. The denormalization we used for follows was fairly mild, however, each follow relationship is stored in exactly two places. For home timelines, we will create a much more aggressively denormalized data structure: a given piece of data will be stored in an arbitrary number of places.

While this highly denormalized structure will be the end result of our work in this chapter, we'll explore several approaches along the way, starting with a fully normalized data structure, proceeding through a partially denormalized approach, and finally settling on a fully denormalized design. We'll discuss the advantages and disadvantages of each, both in terms of implementation complexity and runtime performance. By the end of this chapter, you'll learn:

- How to retrieve data from more than one specific partition
- How to order and paginate in multipartition queries
- How to apply different denormalization strategies to a problem
- How to group multiple write statements into a single operation using logged batches

A normalized approach

Before we proceed down the path of denormalization, let's first try an approach that requires adding no new tables to our schema. Our goal is to display to a user all the status updates of all the users they follow, the most recent first. So, the simplest approach would be to simply look up the followed users, and then retrieve their status updates.

If you've been following along with the code examples so far, you should currently have a single follow relationship in the database, such as alice follows carol. To make things more interesting, let's also have alice follow dave:

```
INSERT INTO "user_outbound_follows"
  ("follower_username", "followed_username")
VALUES ('alice', 'dave');
```

```
INSERT INTO "user_inbound_follows"
  ("followed_username", "follower_username")
VALUES ('dave', 'alice');
```

Now that alice is following a couple of users, we can build a meaningful home timeline for her. Let's have carol and dave write some updates:

```
INSERT INTO "user_status_updates" ("username", "id", "body")
VALUES ('dave', NOW(), 'dave update one');
```

```
INSERT INTO "user_status_updates" ("username", "id", "body")
VALUES ('carol', NOW(), 'carol update one');
```

```
INSERT INTO "user_status_updates" ("username", "id", "body")
VALUES ('dave', NOW(), 'dave update two');
```

```
INSERT INTO "user_status_updates" ("username", "id", "body")
VALUES ('carol', NOW(), 'carol update two');
```

Generating the timeline

We now have all the data we need to generate a home timeline for alice. First, we require a list of users that alice follows:

```
SELECT "followed_username"
FROM "user_outbound_follows"
WHERE "follower_username" = 'alice';
```

This is a straightforward single partition query, simply looking up all of the rows in the `alice` partition in the `user_outbound_follows` table. We'll get back the usernames of those users whose updates `alice` is interested in:

```
          followed_username
          ------------------
                      carol
                       dave
```

```
     (2 rows)
```

Now we need to retrieve all of the status updates for these users:

```
SELECT "id", UNIXTIMESTAMPOF("id"), "body"
FROM "user_status_updates"
WHERE "username" IN ('carol', 'dave');
```

We briefly looked at the WHERE...IN syntax in the *Selecting more than one row* section of *Chapter 2, The First Table*, where we used it to select multiple specific rows from a table with a single column primary key. Here, we generalize this approach to select rows from multiple specific partitions: in this case, we retrieve all the status updates in the `carol` and `dave` partitions. The result is all of the status updates for both users:

```
id                                   | UNIXTIMESTAMPOF(id) | body
-------------------------------------+---------------------+------------------
3a597500-28cf-11e4-8069-5f98e903bf02 |       1408583123024 | carol update one
3a5a3850-28cf-11e4-8069-5f98e903bf02 |       1408583123029 | carol update two
3a58d8c0-28cf-11e4-8069-5f98e903bf02 |       1408583123020 |  dave update one
3a59c320-28cf-11e4-8069-5f98e903bf02 |       1408583123026 |  dave update two
```

One thing we might notice right away is that, while we have the right status updates in the results, they're not in the right order. This is, of course, because results that comprise multiple partitions are returned in the order of the token of the partition key; so we'll see all status updates of `carol` before we see any of `dave`. In the case where the entire timeline comprises only four status updates, it's entirely reasonable to just sort them on the client side.

Ordering and pagination

If our service is even slightly successful, a user's timeline is going to contain thousands or tens of thousands of status updates. We will, of course, want to show only the few most recent by default, and allow the user to paginate through older ones. Fortunately, Cassandra allows us to use the ORDER BY clause in a query that specifies multiple partitions using the IN keyword; under the hood, Cassandra will perform an ordered merge of the rows from the specified partitions.

Let's assume that we only want to show two status updates per page. Accordingly, we'll add the LIMIT and ORDER BY clauses to our query:

```
SELECT "username", "id", UNIXTIMESTAMPOF("id"), "body"
FROM "user_status_updates"
WHERE "username" IN ('carol', 'dave')
ORDER BY "id" DESC
LIMIT 2;
```

Note that, as always, the column given to ORDER BY is the first clustering column; this remains the only valid argument to ORDER BY.

Just as we hoped, we now receive the results in descending order of creation time, regardless of partition key. This is a powerful feature of Cassandra, as it allows us to break out of the narrow constraints of single-partition range queries:

```
username | id                                   | UNIXTIMESTAMPOF(id) | body
---------+--------------------------------------+---------------------+------------------
   carol | 3a5a3850-28cf-11e4-8069-5f98e903bf02 |       1408583123029 | carol update two
    dave | 3a59c320-28cf-11e4-8069-5f98e903bf02 |       1408583123026 | dave update two
```

If we want to retrieve the second page, we can use the same strategy that we used to paginate over slices in a single partition: simply look for the next-newest ID values following the ones we've already seen:

```
SELECT "username", "id", UNIXTIMESTAMPOF("id"), "body"
FROM "user_status_updates"
WHERE "username" IN ('carol', 'dave')
AND "id" < 3a59c320-28cf-11e4-8069-5f98e903bf02
ORDER BY "id" DESC
LIMIT 2;
```

In this query, our WHERE clause plays two roles. As earlier, we specify multiple partitions using the IN keyword but now we also use a range query on id to ensure that we get the second page of results. This is among the most complex WHERE statements that are legal in CQL without resorting to ALLOW FILTERING. As expected, the resulting rows comprise the second page:

```
username | id                                   | UNIXTIMESTAMPOF(id) | body
---------+--------------------------------------+---------------------+------------------
   carol | 3a597500-28cf-11e4-8069-5f98e903bf02 |        1408583123024 | carol update one
    dave | 3a58d8c0-28cf-11e4-8069-5f98e903bf02 |        1408583123020 |  dave update one
```

Multiple partitions and read efficiency

In the real world, most users will likely follow dozens or hundreds of other users. In this case, our WHERE...IN clause will specify hundreds of partitions. Remember from *Chapter 3*, *Organizing Related Data* that each partition is stored separately by Cassandra; querying hundreds of partitions would require hundreds of random accesses. In fact, Cassandra's official documentation warns us against using WHERE... IN in most circumstances:

> "*Under most conditions, using IN in the WHERE clause is not recommended. Using IN can degrade performance because usually many nodes must be queried.*"

Furthermore, in this particular case, Cassandra has to retrieve one page of rows from each partition, perform an ordered merge, and throw away all but the last handful. For instance, if I follow 100 users and have a page size of 10, Cassandra must retrieve 1,000 rows just to figure out which 10 are the most recent. While this approach technically works, its read performance characteristics aren't going to cut it for a production application. We'll need to explore different strategies.

Partial denormalization

Our initial approach to home timelines, which used the existing, fully normalized data structure that we've already built, is technically viable but will perform very poorly at scale. If I follow F users and want a page of size P for my home timeline, Cassandra will need to do the following:

- Query F partitions for P rows each
- Perform an ordered merge of $F \times P$ rows in order to retrieve only the most recent P

The most distressing part of this is the fact that both operations grow in complexity proportionally with the number of people I follow. Let's start by trying to fix this.

The basic goal of the home timeline is to show me the most recent status updates that matter to me. Instead of doing all the work to find out what status updates matter to me, based on who I follow, at read time, let's shift some of the work to write time.

I'll create a table that stores references to status updates that I care about. Whenever someone I follow creates a new status update, I'll add a reference to that update to my home timeline list; this way, when I want to view my home timeline, I'll know exactly what status updates to return.

Let's create our new table:

```
CREATE TABLE "home_status_update_ids" (
  "timeline_username" text,
  "status_update_id" timeuuid,
  "status_update_username" text,
  PRIMARY KEY ("timeline_username", "status_update_id")
)
WITH CLUSTERING ORDER BY ("status_update_id" DESC);
```

Note that there's something subtle going on here with the primary key. In the `user_status_updates` table, the `username` column of the author is the partition key, and the timestamp UUID, `id`, is the clustering column. In order to retrieve a specific row from this table, we need to provide both the `username` and the `id` columns.

However, our `home_status_update_ids` table does not include the author's username in the primary key: the `status_update_username` column is just a normal data column. This is because the `status_update_id` column alone, being a UUID, uniquely identifies the status update, even though it is insufficient on its own to retrieve the status update from the database. We store `status_update_username` as data only so that we know what partition to look in when we go to retrieve the status update.

Now, let's insert some data into our new table. To keep things moving forward, let's ignore the status updates that have already been created; we'll just start using our new timeline structure with new status updates.

Let's say `carol` writes a new status update. Now, along with inserting the status update into her user timeline, we'll need to add a reference to it to all her followers' home timelines. So, the first thing we'll need to do is check who follows her:

```
SELECT "follower_username"
FROM "user_inbound_follows"
WHERE "followed_username" = 'carol';
```

We'll find that `carol`'s only follower is `alice`:

```
          follower_username
         --------------------
                      alice

(1 rows)
```

 While this part of the process might seem simple, we should not do it lightly. Recall that Cassandra is optimized for write performance; because of this, we should strive to avoid reading data during write operations whenever possible. In this case, we are performing a very efficient read operation at write time — retrieving all the rows from a single partition — in order to write a data structure that makes other read operations far more efficient. It's a good tradeoff.

Now we will write all of the data we need to for `carol`'s new status update:

```
INSERT INTO "user_status_updates"
("username", "id", "body")
VALUES (
   'carol', 65cd8320-2ad7-11e4-8069-5f98e903bf02,
   'carol update 3');

INSERT INTO "home_status_update_ids"
   ("timeline_username", "status_update_id",
   "status_update_username")
VALUES
   ('alice', 65cd8320-2ad7-11e4-8069-5f98e903bf02, 'carol');
```

One difference from the previous status update creations jumps out: we're specifying the id column explicitly, instead of using the NOW() function to generate a `timeuuid` on-the-fly. As you'll recall from *Chapter 2, The First Table*, INSERT statements do not give us any feedback: so, if we were to use NOW(), we wouldn't know the UUID that had just been generated. We could try to find out after the fact by reading the most recent id out of `carol`'s partition in the table, but if another process had concurrently created another status update, we'd get the wrong answer.

We need to know with certainty the `id` of the new status update, so that we can use it to create a reference to the status update in `carol`'s followers' home timelines. So, instead of using `NOW()`, we'll generate a UUID in our application. Most languages have perfectly good built-in or third-party libraries for generating UUIDs; in a pinch, you can use a CQL query to do it:

```
SELECT NOW() FROM "user_status_updates" LIMIT 1;
```

Our choice of `user_status_updates` is arbitrary. We could use any non-empty table here, since we're not actually reading any data from the table; rather, we're just exploiting CQL's willingness to generate UUIDs for us, as shown below:

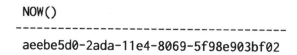

```
NOW()
---------------------------------------
aeebe5d0-2ada-11e4-8069-5f98e903bf02
```

Returning to our earlier status update creation statements, let's examine the second one. Here, we're storing a row that contains three pieces of information: the `username` of a user who is interested in this status update; the `id` of the status update; and the `username` of the author of the status update.

To make things more interesting for `alice`, let's also have `dave` create a new status update:

```
INSERT INTO "user_status_updates"
  ("username", "id", "body")
VALUES
  ('dave', a05b90b0-2ada-11e4-8069-5f98e903bf02, 'dave update 3');
```

```
INSERT INTO "home_status_update_ids"
  ("timeline_username", "status_update_id",
  "status_update_username")
VALUES
  ('alice', a05b90b0-2ada-11e4-8069-5f98e903bf02, 'dave');
```

As with `carol`'s status update earlier, we add a reference to the status update to the home timeline of each of `dave`'s followers, in this case is only `alice`.

Displaying the home timeline

Now, let's say `alice` wants to view her home timeline. In the `home_status_update_ids` table, we have a partition that contains exactly what we need: references to status updates that `alice` is interested in, in the descending order of creation time. So, to know what status updates we want to display to `alice`, we simply perform a range query on her partition:

```
SELECT "status_update_username", "status_update_id"
FROM "home_status_update_ids"
WHERE "timeline_username" = 'alice';
```

Note that we don't need an explicit ORDER BY clause, because we created the table with descending clustering order: as long as we're working within a single partition, results will by default be returned in the order we want.

```
status_update_username | status_update_id
-----------------------+------------------------------------
                  dave | a05b90b0-2ada-11e4-8069-5f98e903bf02
                 carol | 65cd8320-2ad7-11e4-8069-5f98e903bf02

(2 rows)
```

So now we know the primary keys of the status updates we want but we don't have the actual status updates; those live in the `user_status_updates` table. To actually display `alice`'s home timeline, we need to go to that table to retrieve the status updates' content:

```
SELECT * FROM "user_status_updates"
WHERE "username" IN ('dave', 'carol')
AND "id" IN (
   a05b90b0-2ada-11e4-8069-5f98e903bf02,
   65cd8320-2ad7-11e4-8069-5f98e903bf02
);
```

The form that this query takes might be a bit surprising. In reality, we have a collection of compound primary keys—pairs consisting of `username` and `id`—that we would like to retrieve from the table. However, the semantics of the WHERE clause in our query are different: here, we are giving a list of `username` values, and a list of `id` values, and asking for any row that has any arbitrary pair of the given values as a primary key.

This works in the preceding case because the `id` column, as a UUID, is a globally unique identifier for the row. We only specify the `username` column so that Cassandra knows which partitions to go looking for the rows in; as far as identifying the row goes, the `username` is redundant information. If our data structure were such that it were possible for rows with different partition key values to have the same clustering column value, a query of the preceding form would be incorrect.

In this case, though, we get the results we're looking for:

```
 username | id                                   | body
----------+--------------------------------------+----------------
     dave | a05b90b0-2ada-11e4-8069-5f98e903bf02 |  dave update 3
    carol | 65cd8320-2ad7-11e4-8069-5f98e903bf02 | carol update 3

(2 rows)
```

Read performance and write complexity

Let's now examine the performance characteristics of our new data . In the first query, against the `home_status_update_ids` table, we are performing a range query across a single partition: we know that this is an extremely efficient query, in the form we should strive to always use in our application. The second query, however, accesses several partitions, one for each status update author in the result set. If we're using a page size of ten, we might read updates from as many as ten partitions, although it might be fewer if more than one of the updates comes from the same author.

This is an improvement over our previous approach: the number of partitions that we need to access is no longer proportional to the number of followers a user has; it is only proportional to the page size.

The cost of improved read-time performance characteristics is additional write-time complexity. Now, any time a status update is created, the application must determine who follows the author and write a row to the `home_status_update_ids` table for each follower. If the author later gains additional followers, the application might need to go through all of the author's old status updates and add them to the new follower's home timeline.

We also need to ensure that deletions are propagated to the `home_status_update_ids` table. If a status update is deleted, references to it should also be deleted from the home timeline of anyone who follows its author. Let's say that `dave` decides to delete his most recent status update:

```
DELETE FROM "user_status_updates"
WHERE username = 'dave'
```

```
AND id = a05b90b0-2ada-11e4-8069-5f98e903bf02;


DELETE FROM "home_status_update_ids"
WHERE timeline_username IN ('alice')
AND status_update_id = a05b90b0-2ada-11e4-8069-5f98e903bf02;
```

In the second query, we use an IN query even though there is only a single timeline_username in play to demonstrate that this is possible: if dave had many followers, we could simply add more usernames to that clause and still accomplish the deletion with a single query.

In general, the WHERE clause of a DELETE query can take any form that encodes a list of discrete primary key values. Thus the equality and IN operators are allowed; range operators are not. Performing a DELETE without a WHERE clause is also not allowed, as it also fails to specify discrete primary key values.

Fully denormalizing the home timeline

The partially denormalized structure we built using the home_status_update_ids table certainly improves read-time performance, but we're still not at the sweet spot of querying exactly one partition to display the home timeline. In order to do this, we'll need to take the denormalization one step further.

Instead of storing references to status updates in the home timeline, we'll store actual copies of the status updates. Each user's timeline will contain its own copy of the status updates of all the users they follow. We create the following table:

```
CREATE TABLE "home_status_updates" (
  "timeline_username" text,
  "status_update_id" timeuuid,
  "status_update_username" text,
  "body" text,
  PRIMARY KEY ("timeline_username", "status_update_id")
) WITH CLUSTERING ORDER BY ("status_update_id" DESC);
```

This table looks very much like the home_status_update_ids table, except it contains an additional body column. By adding body, we now have a table that will hold a copy of all the data in a given status update; if user_status_updates had ten columns, then we would put all ten columns on home_status_updates too, along with a partition key for the user who owns the timeline.

Creating a status update

Now let's say `carol` and `dave` each create a new status update. As earlier, we will not attempt to backfill `home_status_updates` with status updates that were written prior to its creation; we'll just propagate new status updates to the `home_status_updates` table:

```
SELECT "followed_username"
FROM "user_inbound_follows"
WHERE "followed_username" = 'carol';

INSERT INTO "user_status_updates"
   ("username", "id", "body")
VALUES
   ('carol',
   cacc7de0-2af9-11e4-8069-5f98e903bf02,
   'carol update 4'
);

INSERT INTO "home_status_updates"
   ("timeline_username", "status_update_id",
   "status_update_username", "body")
VALUES (
   'alice',
   cacc7de0-2af9-11e4-8069-5f98e903bf02,
   'carol',
   'carol update 4'
);

SELECT "followed_username"
FROM "user_inbound_follows"
WHERE "followed_username" = 'dave';

INSERT INTO "user_status_updates"
   ("username", "id", "body")
VALUES
   ('dave', 16e2f240-2afa-11e4-8069-5f98e903bf02, 'dave update 4');

INSERT INTO "home_status_updates"
   ("timeline_username", "status_update_id",
   "status_update_username", "body")
VALUES (
   'alice',
   16e2f240-2afa-11e4-8069-5f98e903bf02,
   'dave',
   'dave update 4'
);
```

The flow of the write operation looks pretty much the same as the partially denormalized version:

1. Look up the author's followers.
2. Write the status update to the author's user timeline.
3. Write a copy of the status update to each of the followers' home timelines.

The only difference between partial and full denormalization is that, in the final step, we write *all* the data from the status update to the home timeline, not just the primary key values.

Displaying the home timeline

Now let's display `alice`'s home timeline. All the work we did at write time has paid off: we can now get all the data we need in a single query:

```
SELECT "status_update_username", "status_update_id", "body"
FROM "home_status_updates"
WHERE "timeline_username" = 'alice';
```

Simple as that, we've got `alice`'s home timeline:

```
status_update_username | status_update_id                     | body
-----------------------+--------------------------------------+----------------
                  dave | 16e2f240-2afa-11e4-8069-5f98e903bf02 |  dave update 4
                 carol | cacc7de0-2af9-11e4-8069-5f98e903bf02 | carol update 4

(2 rows)
```

Write complexity and data integrity

The amount of work we need to do to write data in the fully normalized strategy is basically equal to what we needed to do with a partially normalized layout. Our storage needs to increase by a bit, now we're storing one full copy of each status update for every follower the author has. However, storage is cheap, and writing data in Cassandra is cheap, so we've managed to make our timeline read pattern far more efficient at low cost.

One concern in any sort of denormalized scenario is data integrity. At the Cassandra level, the only thing stopping us from adding a status update to the `user_status_updates` table is forgetting to add copies as appropriate to the `home_status_updates` table, or vice versa. Even worse, if a user deletes a status update and we don't properly remove copies from all the `home_status_updates` table, the user's followers might see status updates that they aren't supposed to.

For the most part, the responsibility for maintaining data integrity falls on the application, and there's no magic formula: just well-factored data access logic and lots of tests. However, there is one scenario that is outside the application's control: what if part of the write operation succeeds but another part fails?

In our example, it's possible that, although writing dave's status update to the user_ status_updates table works fine, the write to the home_status_updates table fails because the required nodes are unavailable. By the time the application knows something is wrong, it has already written data to user_status_updates, leaving the data in an invalid state.

One approach would be for the application to delete the row from user_status_ updates if it encounters an error on a subsequent write. However, this sort of manual rollback is error-prone and burdensome, particularly as write operations become more complex.

Happily, Cassandra gives us a cleaner way to ensure that write failures don't lead to a breakdown of data integrity. Multiple write statements can be sent in a single batch; Cassandra will guarantee that, if any statement in a batch succeeds, all of it will. So, we can perform all the write operations to store dave's status update without worry:

```
BEGIN BATCH
  INSERT INTO "user_status_updates"
    ("username", "id", "body")
  VALUES (
    'dave',
    16e2f240-2afa-11e4-8069-5f98e903bf02,
    'dave update 4'
);

  INSERT INTO "home_status_updates" (
    "timeline_username",
    "status_update_id",
    "status_update_username",
    "body")
  VALUES (
    'alice',
    16e2f240-2afa-11e4-8069-5f98e903bf02,
    'dave',
    'dave update 4'
  );
APPLY BATCH;
```

This operation has only two possible outcomes: either it's successful, and rows are written to both `user_status_updates` and `home_status_updates`, or it fails, and no data is written at all. In either case, our data is consistent.

 If you typed the above query into cqlsh, you might have noticed that a new prompt did not appear until after `APPLY BATCH`. This is because, even though it contains multiple write statements, a batch must be sent as a single query to Cassandra. This is in contrast to SQL in which it is valid to open a transaction by sending a `BEGIN TRANSACTION` statement to the database, and to close it with a later statement.

Summary

In this chapter, we explored three different approaches to displaying a user's home timeline. The first approach had the advantage of requiring no additional tables in our database; it was fully normalized. The downside was that reading all the data to display a user's home timeline was prohibitively expensive. The next approach, partial denormalization, kept references to the status updates in each user's home timeline, and allowed us to avoid the worst performance characteristics of the normalized approach. Partial denormalization did not, however, allow us to get to the point where a home timeline could be displayed with only a single query to a single partition.

By storing a full copy of every status update in the home timeline of every follower of the update's author—a fully denormalized approach—we achieved the ideal read-time access pattern—namely, selecting a slice of a single partition in a single table. While read performance increased as we denormalized, write complexity and data integrity concerns also increased. You learned to be vigilant about ensuring that application logic is watertight in keeping denormalized data consistent, and you also learned how to use batches to guarantee that a low-level failure does not leave our data in an inconsistent state.

In these first six chapters, we have explored the Cassandra schema design and data access from a high level, focusing on questions such as

- What tables should my application have?
- How should I structure my primary keys?
- How do I retrieve the data I need?
- When does it make sense to store multiple copies of the same data?

In the next few chapters, we will transition away from these questions and begin to explore the full breadth of Cassandra's functionality. You'll learn how to store collections of data in a single column: how to add new columns to an existing table; how to ensure that an INSERT statement doesn't clobber existing data; and quite a bit more.

7
Expanding Your Data Model

In the preceding chapters, we focused largely on the high-level structure of Cassandra tables, and particularly on the forms and uses of primary keys. Now, we will turn our focus to the data that's stored within tables, exploring advanced techniques to add, change, and remove data.

We created several tables in the MyStatus application, but so far we haven't made any changes to those tables' schemas. In this chapter, we'll introduce the ALTER TABLE statement, which enables us to add and remove columns from the tables in our keyspace.

We'll move on to the UPDATE statement, which is used to change the data in existing rows. You'll learn that INSERT and UPDATE have more in common than our experience with relational databases might lead us to believe, and that INSERT in particular can have unexpected and undesirable effects if not used carefully. We'll also expand our understanding of the DELETE statement, using it to remove data from specific columns in a row, rather than deleting the entire row, as we have so far.

Finally, we'll introduce lightweight transactions as a way to ensure data integrity when modifying data. We'll use transactional inserts to make sure that we don't accidentally overwrite a user record when a different user wants the same username, and transactional updates to avoid writing stale data to existing rows. We'll delve into the ways in which Cassandra's lightweight transactions differ from traditional transactions in a relational database.

By the end of this chapter, you'll know:

- How to view a table's schema in cqlsh
- How to add and remove columns from an existing table
- How to update data in a row
- How to remove the data from a single column

- How to ensure that a newly inserted row doesn't conflict with an existing row's key

- How to perform conditional updates to existing rows

Viewing a table schema in cqlsh

In this chapter, we'll be working with the users table, which we haven't had much interaction with since the early chapters. Before we start making changes to the users table, it would be helpful to have a reminder of what its schema looks like.

One option would be to simply issue a SELECT statement and look at the row headers; however, cqlsh gives us a more elegant way to view the schema, namely, the DESCRIBE TABLE statement:

```
DESCRIBE TABLE "users";
```

The output is a CREATE TABLE statement showing the table's schema as well as all properties for the table:

```
CREATE TABLE users (
  username text,
  email text,
  encrypted_password blob,
  PRIMARY KEY (username)
) WITH
  bloom_filter_fp_chance=0.010000 AND
  caching='KEYS_ONLY' AND
  comment='' AND
  dclocal_read_repair_chance=0.000000 AND
  gc_grace_seconds=864000 AND
  index_interval=128 AND
  read_repair_chance=0.100000 AND
  replicate_on_write='true' AND
  populate_io_cache_on_flush='false' AND
  default_time_to_live=0 AND
  speculative_retry='99.0PERCENTILE' AND
  memtable_flush_period_in_ms=0 AND
  compaction={'class': 'SizeTieredCompactionStrategy'} AND
  compression={'sstable_compression': 'LZ4Compressor'};
```

The part of the output beginning with WITH tells us the table properties for the users table; in this case, the properties are all set to their default values. We can ignore this part of the output as working with table properties goes beyond the scope of this book.

The important part of the output is at the top, listing the columns in the table and telling us which comprise the primary key. We're now reminded that the `users` table has the `username` column as its primary key, and `email` and `encrypted_password` are data columns in the table.

 DESCRIBE is not a core CQL command; if you were to try to use it from a language driver, you would see an error that it is unrecognized. Instead, it is part of an extension to the CQL language that is provided by `cqlsh`. A list of all `cqlsh` extensions to CQL is available at `http://www.datastax.com/documentation/cql/3.1/cql/cql_reference/cqlshCommandsTOC.html`

Adding columns to tables

Let's say we want to allow our users to enter their location in the profile. To store users' location, we need a new column in the `users` table; fortunately, it's perfectly straightforward to add a new column to an existing table:

```
ALTER TABLE "users" ADD "city_state" text;
```

This query instructs Cassandra that we'd like to add a column named `city_state`, of type `text`, to the `users` table. It's identical to the equivalent operation in SQL, although the CQL `ALTER TABLE` statement is much more constrained in the operations it can perform.

Now let's check our table schema again:

```
DESCRIBE TABLE "users";
```

As we hoped, we've got a `city_state` column in the schema. I've omitted the table properties for brevity:

```
CREATE TABLE users (
    username text,
    city_state text,
    email text,
    encrypted_password blob,
    PRIMARY KEY (username)
) WITH
```

Deleting columns

Upon further consideration, we may decide that `location` is a better column name than `city_state`. Cassandra does not allow us to rename existing data columns; however, since we haven't put any data in the `city_state` column yet, we can achieve our goals simply by dropping the `city_state` column and adding a `location` column instead:

```
ALTER TABLE "users" DROP "city_state";
ALTER TABLE "users" ADD "location" text;
```

The `DROP` command within the `ALTER TABLE` statement looks just like the `ADD` command, except that we need not specify the column's type; only its name is sufficient. Looking at the output of `DESCRIBE` again, we've now got the columns set up the way we'd like:

```
CREATE TABLE users (
  username text,
  email text,
  encrypted_password blob,
  location text,
  PRIMARY KEY (username)
) WITH
```

Now that we've got our expanded schema, we can take a look at the actual contents of the table:

```
SELECT * FROM "users";
```

The output now includes our new `location` column:

```
 username | email            | encrypted_password                           | location
----------+------------------+----------------------------------------------+----------
      bob |             null | 0x10920941a69549d33aaee6116ed1f47e19b8e713 |     null
     dave |  dave@gmail.com | 0x6d1d90d92bbab0012270536f286d243729690a5b |     null
    carol | carol@gmail.com | 0xed3d8299b191b59b7008759a104c10af3db6e63a |     null
    alice | alice@gmail.com | 0x8914977ed729792e403da53024c6069a9158b8c4 |     null

(4 rows)
```

Recall from *Chapter 2, The First Table* that Cassandra does not have the concept of
NULL values in the sense that SQL databases do; rather, the null displayed by cqlsh
simply indicates that the rows do not have data in the location column.

 For a full account of the ALTER TABLE command,
refer to the DataStax CQL documentation at http://
www.datastax.com/documentation/cql/3.1/
cql/cql_reference/alter_table_r.html

Updating the existing rows

Now that we've got a new location column, we can add some data to it. Any new
user records we create, of course, can have a location value, but perhaps some
of our existing users would like to input their location value too. To do this, we
need to be able to update existing rows. The process should, once again, look quite
familiar to anyone who is familiar with SQL:

```
UPDATE "users"
SET "location" = 'New York, NY'
WHERE "username" = 'alice';
```

Like the INSERT and DELETE statements, the UPDATE statement does not give us any
feedback on the operation. However, we can confirm it worked by reading from the
users table again:

```
SELECT * FROM "users";
```

As we hoped, alice now has a location:

username	email	encrypted_password	location
bob	null	0x10920941a69549d33aaee6116ed1f47e19b8e713	null
dave	dave@gmail.com	0x6d1d90d92bbab0012270536f286d243729690a5b	null
carol	carol@gmail.com	0xed3d8299b191b59b7008759a104c10af3db6e63a	null
alice	alice@gmail.com	0x8914977ed729792e403da53024c6069a9158b8c4	New York, NY

(4 rows)

 A full reference for the UPDATE statement can be found
in the DataStax CQL documentation at http://www.
datastax.com/documentation/cql/3.1/cql/
cql_reference/update_r.html

Updating multiple columns

As in the SQL UPDATE statement, we can specify multiple column-value pairs to be updated in a single statement. Let's say dave wants to change his email address and also enter a location:

```
UPDATE "users"
SET "email" = 'dave@me.com', "location" = 'San Francisco, CA'
WHERE "username" = 'dave';
```

On checking the table again, we see that both columns have been updated in the dave row:

```
username | email             | encrypted_password                                 | location
---------+-------------------+----------------------------------------------------+------------------
    bob  |            null   | 0x10920941a69549d33aaee6116ed1f47e19b8e713         |             null
   dave  |      dave@me.com  | 0x6d1d90d92bbab0012270536f286d243729690a5b         | San Francisco, CA
  carol  | carol@hotmail.com | 0xed3d8299b191b59b7008759a104c10af3db6e63a         |             null
  alice  |  alice@gmail.com  | 0x8914977ed729792e403da53024c6069a9158b8c4         |     New York, NY
```

Updating multiple rows

As it turns out, bob and carol both live in St. Louis, MO. We could update each of their records individually, but the UPDATE statement also supports the same WHERE... IN construct that we used previously in the SELECT statements. So, we can update both of their records in a single query:

```
UPDATE "users"
SET "location" = 'St. Louis, MO'
WHERE "username" IN ('bob', 'carol');
```

Now, on checking the users table again, we'll see that the change was applied to both rows:

```
username | email             | encrypted_password                                 | location
---------+-------------------+----------------------------------------------------+------------------
    bob  |            null   | 0x10920941a69549d33aaee6116ed1f47e19b8e713         |     St. Louis, MO
   dave  |      dave@me.com  | 0x6d1d90d92bbab0012270536f286d243729690a5b         | San Francisco, CA
  carol  | carol@hotmail.com | 0xed3d8299b191b59b7008759a104c10af3db6e63a         |     St. Louis, MO
  alice  |  alice@gmail.com  | 0x8914977ed729792e403da53024c6069a9158b8c4         |     New York, NY
```

> While the preceding UPDATE statement allows us to update rows in a single statement, these updates are not isolated: it is possible that, while the update is in progress, another client may see a value in one of the rows but not the other. Writes to Cassandra are always isolated within a single partition but there is no way to guarantee isolation across partitions.

Removing a value from a column

Let's say `bob` decides he doesn't want his location on his profile. We'd like to get the `location` column back to the state it was in earlier, which appears as `null` in `cqlsh` but simply means "no data here".

Missing columns in Cassandra

As we discussed previously, Cassandra doesn't have the concept of `NULL` columns in the SQL sense. Concretely, the following statement is not possible in Cassandra:

```
SELECT * FROM "users" WHERE "location" IS NULL;
```

Relational databases typically store a separate bit in each column to indicate whether that column contains a `NULL` value. In Cassandra, on the other hand, all data columns are optional, and only the columns with a value are represented in storage. We can visualize Cassandra rows as maps of key-value pairs, with some keys possibly missing; relational database rows are more like fixed size lists of columns, where some columns may have a `NULL` value. The following diagram shows how we might visualize `dave`'s row in Cassandra, and the equivalent row in a relational database:

Cassandra

username	email	encrypted_password
dave	dave@gmail.com	0x6d1d90d

Relational Database

username	email	encrypted_password	location
dave	dave@gmail.com	0x6d1d90d	NULL

Deleting specific columns

To remove data from a column in Cassandra, rather than setting it to `NULL`, we'll want to simply delete it from the column. The syntax for CQL's `DELETE` statement allows us to do just this by specifying one or more specific columns that we want to delete from:

```
DELETE "location"
FROM "users"
WHERE "username" = 'bob';
```

We can now check the `users` table again to confirm that `bob` no longer has a location in his profile:

```
username | email              | encrypted_password                             | location
---------+--------------------+------------------------------------------------+--------------------
     bob |               null | 0x10920941a69549d33aaee6116ed1f47e19b8e713 |                null
    dave |        dave@me.com | 0x6d1d90d92bbab0012270536f286d243729690a5b | San Francisco, CA
   carol | carol@hotmail.com  | 0xed3d8299b191b59b7008759a104c10af3db6e63a |       St. Louis, MO
   alice |    alice@gmail.com | 0x8914977ed729792e403da53024c6069a9158b8c4 |       New York, NY
```

Syntactic sugar for deletion

Although Cassandra doesn't have real NULL columns, CQL does provide some syntactic sugar to provide a more familiar interface to remove data from columns. You can, in fact, write UPDATE statements that set values to `null`:

```
UPDATE "users"
SET "location" = NULL
WHERE "username" = 'alice';
```

Note that this is purely syntactic sugar: the preceding statement is precisely the same operation as the following one:

```
DELETE "location"
FROM "users"
WHERE "username" = 'alice';
```

The DELETE form of the statement is more expressive of what we are actually doing—removing a value from a column—but the UPDATE form is more familiar for anyone used to working with with SQL. Another advantage of the UPDATE statement is that we can write a statement that simultaneously sets data in some columns, and removes data from others:

```
UPDATE "users"
SET "email" = 'carol@hotmail.com', "location" = NULL
WHERE username = 'carol';
```

Checking up on the status of the `users` table, we can see that `carol`'s row reflects our latest changes:

```
username | email              | encrypted_password                             | location
---------+--------------------+------------------------------------------------+--------------------
     bob |               null | 0x10920941a69549d33aaee6116ed1f47e19b8e713 |                null
    dave |        dave@me.com | 0x6d1d90d92bbab0012270536f286d243729690a5b | San Francisco, CA
   carol | carol@hotmail.com  | 0xed3d8299b191b59b7008759a104c10af3db6e63a |                null
   alice |    alice@gmail.com | 0x8914977ed729792e403da53024c6069a9158b8c4 |                null
```

Inserts, updates, and upserts

So far, we've used the INSERT statements to add new rows to our tables, and the UPDATE statements to update information in existing rows. As it turns out, both INSERT and UPDATE statements can modify existing rows and can create new rows. At their core, we can most accurately think of the INSERT and UPDATE statements as providing different syntax for the same underlying operation, an **upsert**.

This is quite astonishing for those of us who are used to SQL, in which the INSERT and UPDATE statements are entirely distinct. While there are some situations in which upsert behavior is quite handy, it can also be a stumbling block, especially for developers who are new to Cassandra. Fortunately, Cassandra offers us ways to ensure that our write operations behave the way we intend; we'll explore these techniques after taking a closer look at upserts.

Inserts can overwrite existing data

Let's add a fifth user to our application, eve. We'll sign her up with a full complement of profile fields:

```
INSERT INTO "users"
("username", "email", "encrypted_password", "location")
VALUES
('eve', 'eve@gmail.com',
   0x85e36ed9f726295921a4a6f40d95c202c895180c,
   'Washington, D.C.');
```

On checking the users table, we can confirm that eve has the data we intended:

```
username | email           | encrypted_password                         | location
---------+-----------------+--------------------------------------------+------------------
    bob  |            null | 0x10920941a69549d33aaee6116ed1f47e19b8e713 |             null
    dave |     dave@me.com | 0x6d1d90d92bbab0012270536f286d243729690a5b | San Francisco, CA
    eve  |   eve@gmail.com | 0x85e36ed9f726295921a4a6f40d95c202c895180c | Washington, D.C.
   carol | carol@hotmail.com | 0xed3d8299b191b59b7008759a104c10af3db6e63a |             null
   alice |  alice@gmail.com | 0x8914977ed729792e403da53024c6069a9158b8c4 |             null
```

Now let's imagine that another user, also named eve, would like to use MyStatus. If we naïvely allow her to choose any username she wants, we will end up attempting to create her username like this:

```
INSERT INTO "users"
("username", "email", "encrypted_password")
VALUES
('eve', 'eve123@hotmail.com',
   0x411ebd49a24f8be2250c527160509264622d7883);
```

Intuitively, we would expect this statement to generate an error because we're trying to insert a row with the primary key eve, which is already assigned to an existing row. In fact, Cassandra will report no such error and, much to our horror, we'll find that we've overwritten data from the first eve's profile:

username	email	encrypted_password	location
bob	null	0x10920941a69549d33aaee6116ed1f47e19b8e713	null
dave	dave@me.com	0x6d1d90d92bbab0012270536f286d243729690a5b	San Francisco, CA
eve	eve123@hotmail.com	0x411ebd49a24f8be2250c527160509264622d7883	Washington, D.C.
carol	carol@hotmail.com	0xed3d8299b191b59b7008759a104c10af3db6e63a	null
alice	alice@gmail.com	0x8914977ed729792e403da53024c6069a9158b8c4	null

Looking at the current state of the users table, we can see that the email and encrypted_password columns from the original eve's account have been overwritten by the new eve. However, the location column is untouched—it still shows Washington, D.C., the value from the first eve's account.

This is because we did not specify a location column in the second INSERT statement. The INSERT statements only touch columns that are listed in the parenthesized list of columns that directly follows the table name. So, we can think of our most recent query as saying, "set the email and encrypted_password columns in the row with key eve to the specified values."

Clearly, we've got a problem. The row at key, eve, now contains a mishmash of information from two different user profiles, and we've lost data from the first eve's profile. How can we prevent this from happening?

Checking before inserting isn't enough

Our first instinct might simply be to check the users table to make sure there isn't already a row with the requested username in the table. To keep things concise, we might do something like the following:

```
SELECT "username" FROM "users" WHERE "username" = 'eve' LIMIT 1;
```

This will return a single row containing only the username value, if the username value is taken:

```
username
----------
eve

(1 rows)
```

If the username is available, no rows will be returned. While checking the availability of a username is a good way to give the user early feedback that they need to choose a different name, it's not enough to guarantee that we won't accidentally overwrite an existing row. The reason is that this sequence of operations is subject to a race condition in the case where two users are trying to sign up with the same name at roughly the same time. Concretely, we will see the following sequence of events, with both "eve 1" and "eve 2" trying to create an account with username eve at the same time:

1. The character eve 1 sends a request to our application to create an account.

2. The character eve 2 sends a request to our application to create an account.

3. The process creating eve 1's account checks for an existing row with username eve. It finds that there are none.

4. The process creating eve 2's account checks for an existing row with username eve. It also finds that there are none.

5. The process creating eve 1's account issues an INSERT statement to create a new row with username, eve.

6. The process creating eve 2's account issues an INSERT statement to create a row with username, eve, overwriting the row created for eve 1.

For the problem to occur, Step 4 must happen before Step 5: it's important that both processes check the database for an existing row before either writes to the database. It's an improbable scenario, in which two users would attempt to sign up with the exact same username at roughly the exact same time, but it's not an impossible one. As our service becomes more popular, we're more and more likely to run across such unlikely events, so we're much better off with a system that guarantees that no new user account will overwrite an existing one.

Another advantage of UUIDs

As it turns out, the users table is the only table in the MyStatus keyspace that has this problem. All the other tables use UUIDs in their primary keys; since UUIDs are guaranteed to be globally unique, we simply don't have to worry about what might happen in the case of a primary key collision.

UUIDs are great in the case where the table doesn't have a natural key, as we discussed in *Chapter 2, The First Table*. However, using a UUID key in the users table won't solve the fundamental problem, since we still want to guarantee that usernames are unique. A username collision would be less catastrophic, since at least we wouldn't overwrite data; However, we'd have also forfeited the only structure Cassandra gives us for guaranteeing uniqueness, the primary key. Fortunately, there's a better way.

Conditional inserts and lightweight transactions

Version 3.1 of CQL introduced new support for **lightweight transactions**, which allow us to modify data only if certain conditions are met. Unlike the application-level integrity checks we discussed earlier, lightweight transactions are safe in concurrent environments: if two processes attempt to conditionally create a row with the same key, lightweight transactions guarantee that only one will succeed.

To perform a conditional insert, we simply add the clause, `IF NOT EXISTS`, to the end of the `INSERT` statement. Let's start fresh with a new user account, `frank`:

```
INSERT INTO "users"
("username", "email", "encrypted_password", "location")
VALUES
('frank', 'frank@gmail.com',
  0xa71451665e16d8c6e6edfd444c60156efc861432,
  'Los Angeles, CA')
IF NOT EXISTS;
```

When we perform this insert, we'll immediately see something surprising. Cassandra has feedback on the operation:

```
 [applied]
-----------
   True
```

While normal write operations give no feedback on their outcome, lightweight transactions do give feedback. After all, a conditional insert would only be applied if the row did not exist prior to the write operation; we would like to know what the outcome was. To see how conditional inserts behave with a key collision, we can attempt to insert another user with the username `frank`:

```
INSERT INTO "users"
("username", "email", "encrypted_password")
VALUES
('frank', 'frank123@hotmail.com',
  0xed945334b49ea005967cd4c48ff0493e2d3ff8d3)
IF NOT EXISTS;
```

We'll expect this insert to *not* be applied, since we are performing a conditional insert and there is already a row with the username `frank`. Indeed, Cassandra's feedback for us is that the insert did not happen:

```
[applied] | username | email          | encrypted_password                         | location
----------+----------+----------------+--------------------------------------------+----------------
   False  |    frank | frank@gmail.com | 0xa71451665e16d8c6e6edfd444c60156efc861432 | Los Angeles, CA
```

As with the feedback for a successful application, we see an `[applied]` column, which tells us the outcome of the lightweight transaction. In the case where a key collision prevented it from being applied, Cassandra also helpfully returns the data in the existing row with the key we were trying to write to.

> For a detailed account of how lightweight transactions are implemented, refer to the DataStax Cassandra documentation page at `http://www.datastax.com/documentation/cassandra/2.1/cassandra/dml/dml_ltwt_transaction_c.html`

Conditional inserts allow us to protect the data integrity of tables that use a natural primary key that is not globally unique. However, overwriting existing rows isn't the only way Cassandra's upsert behavior can get us into trouble.

Updates can create new rows

The `INSERT` statements aren't the only write operations in CQL that can have unexpected effects. To see how `UPDATE` can take us by surprise, let's update `gina`'s row to add a location:

```
UPDATE "users"
SET "location" = 'Houston, TX'
WHERE "username" = 'gina';
```

Wait a second—who's `gina`? So far, we haven't had a user with the username `gina` in our table. We might expect the preceding update statement to have no effect: after all, our `WHERE` statement specifies a row that does not exist.

The contents of the `users` table, however, tell a different story:

```
username | email               | encrypted_password                                | location
---------+---------------------+---------------------------------------------------+-------------------
     bob |                null | 0x10920941a69549d33aaee6116ed1f47e19b8e713        |              null
    dave |         dave@me.com | 0x6d1d90d92bbab0012270536f286d243729690a5b        | San Francisco, CA
    gina |                null |                                                   |        Houston, TX
     eve | eve123@hotmail.com  | 0x411ebd49a24f8be2250c527160509264622d7883        | Washington, D.C.
   carol |   carol@hotmail.com | 0xed3d8299b191b59b7008759a104c10af3db6e63a        |              null
   frank |      frank@gmail.com | 0xa71451665e16d8c6e6edfd444c60156efc861432       | Los Angeles, CA
   alice |     alice@gmail.com | 0x8914977ed729792e403da53024c6069a9158b8c4        |              null
```

There is now a row with the primary key `gina`. It turns out that an UPDATE statement, just like an INSERT statement, is in fact an upsert operation: if the specified row exists, it will update values in the row, but if it doesn't, a new row will be created.

As it turns out, in their basic forms, the INSERT and UPDATE queries are syntactic variants of the exact same underlying operation. For instance, the following two queries are functionally identical to one another:

```
INSERT INTO "users"
("username", "email", "encrypted_password")
VALUES
('alice', 'alice@gmail.com',
   0x8914977ed729792e403da53024c6069a9158b8c4);

UPDATE "users"
SET "email" = 'alice@gmail.com',
"encrypted_password" =  0x8914977ed729792e403da53024c6069a9158b8c4
WHERE "username" = 'alice';
```

In both cases, the query simply instructs Cassandra to set the `email` and `encrypted_password` columns to the given values in the row whose primary key is `alice`.

INSERT and UPDATE are not entirely redundant; however, each provides certain capabilities that the other does not. For instance, only INSERT supports the IF NOT EXISTS clause that we explored earlier. UPDATE queries also have distinctive functionality, not the least of which is conditional updates.

Optimistic locking with conditional updates

As the MyStatus service becomes more popular, we're likely to see organizations beginning to create accounts that are managed by multiple people. Once we have multiple people accessing the same resource at the same time, we need to think about what might go wrong when different people try to make conflicting changes to the same piece of data.

Let's create a new account on behalf of an organization called HappyCorp. We'll just start with some basic information:

```
INSERT INTO "users"
("username", "email", "encrypted_password")
VALUES
('happycorp', 'media@happycorp.com',
  0x368200fa910c16cc644f3512e63b541c85fa2a3c)
IF NOT EXISTS;
```

We use a conditional update, of course, to ensure that we're not overwriting another row with the same primary key.

Now, let's imagine that two members of the HappyCorp social media team decide to update the location listed in the company's profile on MyStatus at the same time. One of them elects to enter New York, the location of HappyCorp's corporate offices, and the other thinks it would make more sense to list Palto Alto, where the company's main R&D campus is located. Unfortunately, the two employees decide to make this update at the exact same time; so each one sees a blank location field, types a value in, and tries to save it without realizing that they are stomping on each other's work.

A popular solution to this sort of problem is known as **optimistic locking**, which uses a version column to ensure that two simultaneous updates will not conflict with each other. Each time a record is displayed to a user for the purpose of updating it, the application keeps track of what version was displayed (on the web, this might be done using a hidden form field). Changes are saved to the database using a **conditional update**, ensuring that they're only applied if the *current* version in the database is the same as the version that was initially displayed to the user. As part of the operation, we also update the version itself.

To see how this works, let's first add a version column to the users table. We could do this with a simple integer, incrementing it each time we update a row; however, given Cassandra's excellent support for UUIDs, we might as well use one and get the added advantage of knowing when a column was last updated:

```
ALTER TABLE "users" ADD "version" timeuuid;
```

Now we need to give each existing row a version. This can be done easily enough with a WHERE...IN update query:

```
UPDATE "users"
SET "version" = NOW()
WHERE "username" IN (
  'alice', 'bob', 'carol', 'dave', 'eve', 'frank', 'gina',
  'happycorp'
);
```

This will give every user profile the same initial version UUID, but that's fine; we just need to make sure version numbers are unique over time within a single row.

Optimistic locking in action

Now it's time to implement actual optimistic locking. When we initially display the profile to the two HappyCorp employees for editing, we'll keep track of the version that we displayed. Then, when the first employee goes to save their update, we'll perform a conditional update to make sure that we're updating the same version that the user thinks we are:

```
UPDATE "users"
SET "location" = 'New York, NY', "version" = NOW()
WHERE "username" = 'happycorp'
IF "version" = ec0c1fb7-321f-11e4-8eeb-5f98e903bf02;
```

The IF clause at the end of the query makes this a conditional update, another type of lightweight transaction. You'll notice that, just like conditional inserts, conditional updates give us feedback on whether the update was applied:

```
 [applied]
-----------
   True
```

The key fact about this optimistically locked update is that we do two things with the version column. First, we use a conditional update to instruct Cassandra to only perform the update if the version is the same as it was when the user initiated the update. Second, in the same query, we change the version column to a new value. Taken together, these two interactions with the version column ensure that we cannot unwillingly overwrite changes made by another user to the same row.

On checking the relevant columns in the users table, we see that both the location and version columns have been updated:

```
 username  | location     | version
-----------+--------------+-------------------------------------
 happycorp | New York, NY | 8a810110-3220-11e4-8eeb-5f98e903bf02
```

Now, let's see how our system behaves in the case of concurrent updates by different users. The situation we're most concerned about is the one in which a second user of the system also tries to update HappyCorp's location, and initiates the process *before* the update we just performed is completed. From the perspective of this second updater, the location field is blank, and we know that the version we've read is the original version, ec0c1fb1-321f-11e4-8eeb-5f98e903bf02.

So, when the second updater tries to save their work, we'll once again make the update conditional on finding the version number we started with:

```
UPDATE "users"
SET "location" = 'Palo Alto, CA', "version" = NOW()
WHERE "username" = 'happycorp'
IF "version" = ec0c1fb7-321f-11e4-8eeb-5f98e903bf02;
```

From a bird's-eye view of this process, we know that this update will not be applied because the process that updated the `location` value to `New York` also changed the `version`. However, the process that's trying to make the change to `Palo Alto` doesn't need to know this fact; it can simply ask Cassandra to make the update conditional on finding the right version, and Cassandra will do the right thing no matter what. In this case, we see that the update is indeed not applied:

```
[applied] | version
----------+-------------------------------------
    False | 8a810110-3220-11e4-8eeb-5f98e903bf02
```

Much like conditional inserts, conditional updates include relevant information from the row in question if the update is not applied. In this case, it tells us what the actual version in the row is; we see that it's the one that we set when we changed the location to New York.

Optimistic locking and accidental updates

Our optimistic locking strategy also solves another problem: that of accidentally creating or recreating rows when we think we're updating existing rows. By only allowing the update to happen if the `version` column matches what we expect it to, we also implicitly ensure that the row exists. For instance, let's see what happens if we try to modify a nonexistent row with a conditional update:

```
UPDATE "users"
SET "location" = 'Denver, CO', "version" = NOW()
WHERE "username" = 'ivan'
IF "version" = ec0c1fb7-321f-11e4-8eeb-5f98e903bf02;
```

Happily, the update is not applied, so we don't unwillingly create a new row for this `ivan` character:

```
[applied]
----------
    False
```

Since there's no row to take a `version` value, Cassandra omits that information from the feedback, simply telling us that the operation was not applied.

If you're not using optimistic locking, you can get the same effect by adding a `boolean` column that is set to `true` for every row in the table, and performing conditional updates based on the presence of `true` in that column.

Lightweight transactions have a cost

Lightweight transactions allow us to maintain data integrity in the face of concurrent updates but they don't do it for free. Because of Cassandra's distributed architecture, it's actually quite involved to guarantee that the data is in a certain state before modifying it, as all the machines that store that piece of data need to be in agreement. Accordingly, there's a performance penalty in using lightweight transactions; thus, you shouldn't use them in situations where you don't need to.

When lightweight transactions aren't necessary

As we discussed earlier, concurrent insertions aren't a concern for any table that uses UUIDs that are generated by the application or by Cassandra at row creation time. Simply using UUIDs guarantees we'll never have a row key collision.

Another scenario in which we can skip conditional inserts is when we have a globally unique natural key. For instance, if you're building an **RSS** (short for **Rich Site Summary**) reader, you might identify individual posts by their canonical URL. There's no concern about overwriting one post with another post's data because each row has a key that acts as a global unique identifier. In that case, you may choose to periodically update a local copy of the feed by downloading the latest posts from the RSS server.

It truly doesn't matter whether any given post already exists in your local table or not, since you'll only be overwriting data with the same data. This is a situation in which Cassandra's upsert behavior allows us to perform very efficient write-without-read updates.

Summary

In this chapter, we added several new tools to our arsenal of Cassandra knowledge. You now know how to add and remove columns from existing tables in our database schema, and we explored the UPDATE query. You learned that CQL's INSERT and UPDATE have behavior that's quite surprising to those who are accustomed to the equivalent operations in SQL, and that in reality both queries in CQL are performing an underlying "upsert" operation. We explored situations in which upserts can produce undesirable behavior, particularly in situations where there is concurrent access to the same resource by multiple actors. Also, we introduced lightweight transactions in the form of conditional inserts and conditional updates, as a way to mutate Cassandra data with some of the data integrity guarantees that we are used to having in relational databases.

In the next chapter, we will undertake an exploration of data structures *within* individual columns, exploring several different ways to store and manipulate multiple values within a single column.

8

Collections, Tuples, and User-defined Types

In the preceding chapters, we focused largely on the high-level structure of Cassandra tables, and particularly on the forms and uses of primary keys. We paid scant attention to data columns, generally defining our table schemas with one or two text columns at most. While this type of scalar data column is the workhorse of Cassandra schemas, they're not the only arrow in Cassandra's quiver.

Cassandra offers three different structures that store multiple values in a single column. Collection columns can store an arbitrary number of values, all of the same type. Tuples and user-defined types store a fixed number of values but these values can be of different types. While the fields in a tuple are specified by position, user-defined types give a name to each field; when reading data from a user-defined type, it's possible to retrieve a subset of the fields within the column.

Collections come in three flavors: lists, sets, and maps. Set columns store unordered, unique collections of values. List columns store ordered collections of values, and allow duplicates. Map columns store dictionaries of key-value pairs.

The true power of collection columns comes from the ability to make discrete modifications to their contents, without reading their value first. It is possible, for instance, to push a value on to the end of a list, without reading the list's current contents first. We will see many more such examples in this chapter.

Collection columns are particularly useful in scenarios with multiple processes concurrently writing data to the same target. However, collection columns have their limitations, and are not appropriate for every scenario in which they could conceivably be used. We'll discuss the upsides and downsides of collection columns in this chapter.

Tuples and user-defined types are useful when we want to store a single piece of data that has a complex structure. They are especially powerful when used in combination with other CQL features, particularly secondary indexes. We'll see how tuples and user-defined types can be used to work around one of the major limitations of secondary indexes in Cassandra: the restriction of indexes to a single column.

By the end of this chapter, you'll know:

- How to declare a collection column
- The different types of collection columns
- How to add and remove values from a collection
- How to create secondary indexes on collections
- The limitations of collection columns
- How to declare a tuple column
- How to create a user-defined type
- How to declare a column with a user-defined type
- How tuples and user-defined types facilitate more powerful secondary indexes

The problem with concurrent updates

Let's add a new feature to our My Status application: the users' ability to "star" their friends' status updates to indicate their approval. For each status update, we'll store a list of the users who have starred that status update so that we can display the usernames to the author of the update.

Serializing the collection

One approach is to simply store the list of users in a text column in some serialized form. JSON is a versatile serialization format for such scenarios, so we'll use it. First, we'll add the column to the user_status_updates table, recalling the technique from *Adding columns to tables* section in *Chapter 7, Expanding Your Data Model*:

```
ALTER TABLE "user_status_updates"
ADD "starred_by_users" text;
```

Simple enough. Now let's suppose that bob wants to star one of the status updates of alice. From the application's standpoint, this means appending bob to the list of users who have starred alice's update. First, we'll read the existing list:

```
SELECT "starred_by_users"
FROM "user_status_updates"
WHERE "username" = 'alice'
AND "id" = 76e7a4d0-e796-11e3-90ce-5f98e903bf02;
```

We see that there's currently no value in the starred_by_users column, which we can just interpret as an empty list:

```
          starred_by_users
          ------------------
                      null

(1 rows)
```

Since the column was empty before, we know that the new value of the column should just be a one-element array containing bob. We can update the status update row accordingly:

```
UPDATE "user_status_updates"
SET "starred_by_users" = '["bob"]'
WHERE "username" = 'alice'
AND "id" = 76e7a4d0-e796-11e3-90ce-5f98e903bf02;
```

Note that the value in the CQL above is simply a string literal containing a JSON representation of the array ["bob"]. From Cassandra's standpoint, it is simply text.

Introducing concurrency

While this approach works fine in isolation, we'll run into a problem if more than one process needs to concurrently update the starred_by_users column in the same row. Concretely, if two users star a status update at roughly the same time, we'll run into a problem. Consider the following sequence of events:

1. Both carol and dave want to star alice's status update.

2. Each of their requests is handled by a separate, concurrent process within the MyStatus application.

3. The process of the character carol receives the request and begins by reading the current value of the starred_by_users column, retrieving ["bob"].

4. The process of the character dave receives its request, and also reads the current value of starred_by_users, which is still ["bob"].

5. The process of the character `carol` adds her name to the list and writes it back to the row, setting it to `["bob", "carol"]`.

6. The process of the character `dave` adds his name to the list that it read, oblivious to the fact that it has now changed in the database, writing `["bob", "dave"]` to the row.

7. The character `carol` has now lost the star of the status update.

This is a similar scenario to the one we explored in *Chapter 7, Expanding Your Data Model* with concurrent updates to the `users` table and, indeed, optimistic locking would be a viable solution. By introducing a version column to the `user_status_updates` table and performing conditional updates, we could ensure that the above scenario did not result in data loss. In particular, step 6 would result in a conditional update not applied. The process of the character `dave` could then reread the updated value of the `starred_by_users` list, append the username of `dave` to the list, and retry the update.

This solution works but it's not completely satisfying. In particular, it requires at least one read operation from the database before the row can be updated; in the case of a version conflict, additional reads are required. As we discussed in the previous chapter, conditional writes do carry a performance cost. Finally, this approach adds additional complexity to our application: we need to have code that handles the case of a stale version and retries the operation.

Collection columns and concurrent updates

Happily, Cassandra's collection columns allow us to bypass the entire read-mutate-write process, allowing us to express the exact operation we want to do in CQL without regard to the current value of the column.

Defining collection columns

Given the limitations of the serialized approach, let's drop our `text` column and replace it with a `collection` column of the same name. Cassandra offers three flavors of collections: lists, sets, and maps. We'll explore all three in this chapter, starting with a set column, which is the most appropriate for our starred_by_users column:

```
ALTER TABLE "user_status_updates"
DROP "starred_by_users";

ALTER TABLE "user_status_updates"
ADD "starred_by_users" SET<text>;
```

The last line introduces a new syntax to define a collection column type. The new column that we have created is a collection column using the SET data structure; the values it contains have type text. Like normal scalar data columns, collection columns must define the type of their values, and any type is permitted.

As the name suggests, set columns contain collections of values with the following properties:

- Values are not in a defined order
- Values are unique within the collection

This is just what we want for our starred_by_users column: a user can only star a status update once, and we don't care about the order in which users starred an update.

Reading and writing sets

Before getting into the more powerful operations we can perform on sets, let's start with the basics. Rerunning the scenario we explored earlier, let's naïvely use the exact same approach as we did the first time to record bob's status update. First, we'll read the value of the set:

```
SELECT "starred_by_users"
FROM "user_status_updates"
WHERE "username" = 'alice'
AND "id" = 76e7a4d0-e796-11e3-90ce-5f98e903bf02;
```

Once again, we see that the value of the set is currently null:

```
         starred_by_users
--------------------------------------
                     null


(1 rows)
```

This means that we can write a one-element set containing the text value bob to the starred_by_users column in the status update:

```
UPDATE "user_status_updates"
SET "starred_by_users" = {'bob'}
WHERE "username" = 'alice'
AND "id" = 76e7a4d0-e796-11e3-90ce-5f98e903bf02;
```

This query introduces some new syntax, specifically the value we're writing to `starred_by_users`. The value we specify is a set literal, which is simply a comma-delimited list of values surrounded by curly braces. Each collection type has its own literal format.

Let's check the contents of the status update row to see the effect of our update:

```
 username | id                                   | starred_by_users
----------+--------------------------------------+------------------
    alice | 76e7a4d0-e796-11e3-90ce-5f98e903bf02 |           {'bob'}

(1 rows)
```

Observe that `cqlsh` outputs the value of a set column using the same set literal we encountered previously, and that our update of the column worked as intended.

Advanced set manipulation

This access pattern is quite familiar, and we can anticipate that it's just as vulnerable to data loss as the serialized approach we explored previously. The beauty of a collection column is that, if we want to add a value to the set, we can write a CQL statement to do just that. Let's pick up our scenario with the addition of `carol`'s star to `alice`'s status update:

```
UPDATE "user_status_updates"
SET "starred_by_users" = "starred_by_users" + {'carol'}
WHERE "username" = 'alice'
AND "id" = 76e7a4d0-e796-11e3-90ce-5f98e903bf02;
```

Note that, in the above statement, the previous value of the `starred_by_users` column does not appear; in fact, during the entire operation, there is no need for us to read data from the row at all. The SET clause in our UPDATE statement simply instructs Cassandra to set `starred_by_users` to the union of its current value and the one-element set containing `'carol'`; or, put more simply, to add `'carol'` to the existing set.

Let's check the value of the set after our update:

```
 username | id                                   | starred_by_users
----------+--------------------------------------+------------------
    alice | 76e7a4d0-e796-11e3-90ce-5f98e903bf02 | {'bob', 'carol'}

(1 rows)
```

As we instructed, the set now contains both `'bob'` and `'carol'`, putting it in the exact state we would expect. In similar fashion, we can now add `'dave'` to the list:

```
UPDATE "user_status_updates"
SET "starred_by_users" = "starred_by_users" + {'dave'}
WHERE "username" = 'alice'
AND "id" = 76e7a4d0-e796-11e3-90ce-5f98e903bf02;
```

Again, on reading the row, we can confirm that the list contains all the usernames that we expected:

```
username | id                                    | starred_by_users
---------+---------------------------------------+-------------------------
   alice | 76e7a4d0-e796-11e3-90ce-5f98e903bf02  | {'bob', 'carol', 'dave'}
```

(1 rows)

Crucially, this approach is entirely resilient to the concurrent update scenario we explored previously. The problem with concurrent updates in the naïve approach was that one process read a value for starred_by_users that subsequently became stale when a different process wrote an updated value to the column. This problem simply can't occur with our strategy of directly adding a new value to the set in CQL, for the simple reason that we never actually read a value that can become stale. The underlying structure of Cassandra sets guarantees that concurrent additions to the set will not result in data loss; the new value is added discretely to the set, not as part of a full overwrite of the set's values.

Removing values from a set

Let's say that dave has second thoughts and no longer wishes to have starred alice's status update. We can just as easily remove his name from the list:

```
UPDATE "user_status_updates"
SET "starred_by_users" = "starred_by_users" - {'dave'}
WHERE "username" = 'alice'
AND "id" = 76e7a4d0-e796-11e3-90ce-5f98e903bf02;
```

This is identical to our previous query, except that we use a – operator to indicate removal rather than a + operator to indicate addition.

Sets and uniqueness

What happens if we try to add a value to a set that it already contains? Let's try adding `carol` to the list again:

```
UPDATE "user_status_updates"
SET "starred_by_users" = "starred_by_users" + {'carol'}
WHERE "username" = 'alice'
AND "id" = 76e7a4d0-e796-11e3-90ce-5f98e903bf02;
```

Now, let's read the row from the database to see our update's effects:

```
 username | id                                   | starred_by_users
----------+--------------------------------------+------------------
    alice | 76e7a4d0-e796-11e3-90ce-5f98e903bf02 | {'bob', 'carol'}

(1 rows)
```

As the definition of the SET data structure suggests, we see only one instance of `carol` in the set; the column itself guarantees the uniqueness of its values.

Collections and upserts

In our previous example, we directly set the initial value of the `starred_by_users` column in the status update row, and only then performed discrete additions and removals. What happens if we try to add a value to a set that doesn't exist? Let's try it on a different status update, one that has no value in the `starred_by_users` column:

```
UPDATE "user_status_updates"
SET "starred_by_users" = "starred_by_users" + {'alice'}
WHERE username = 'bob'
AND id = 97719c50-e797-11e3-90ce-5f98e903bf02;
```

On running the query, we don't get any error message, which is a promising sign. Let's have a look at the contents of that row now:

```
 username | id                                   | starred_by_users
----------+--------------------------------------+------------------
      bob | 97719c50-e797-11e3-90ce-5f98e903bf02 |        {'alice'}

(1 rows)
```

As it turns out, we now have a one-element set containing alice's username. Adding data to a set is, in fact, an upsert operation: the collection will be created as part of the operation if it does not already exist. This observation generalizes to all operations on all collection column types; you never need to check for the existence of a collection before adding data to it.

Using lists for ordered, nonunique values

Let's say that we'd like to allow users to share other users' status updates with their own followers. For each status update, we'll keep track of the shares, so that we can display them to the author. Unlike starring a status update, the same user can share the same status update more than once, so we will track shares as discrete events, retaining the order in which they occurred.

A list column fits this task perfectly. Like sets, lists contain collections of values; but, unlike sets, values in sets are nonunique and have a stable and defined order.

Defining a list column

First, let's add a new list column to our user_status_updates table:

```
ALTER TABLE "user_status_updates"
ADD "shared_by" LIST<text>;
```

The syntax for defining a list column is identical to that for defining a set column; we simply swap in LIST for SET.

Writing a list

As with sets, we can directly specify the entire contents of a list, overwriting the current contents:

```
UPDATE "user_status_updates"
SET "shared_by" = ['bob']
WHERE "username" = 'alice'
AND "id" = 76e7a4d0-e796-11e3-90ce-5f98e903bf02;
```

The literal for a list is just like that of a set, except that square brackets are used instead of curly braces. This is, of course, the syntax for array literals in a wide variety of programming languages.

Discrete list manipulation

Lists, like sets, allow us to discretely add and remove elements without reading the existing contents or overwriting the entire collection. However, since elements in a list have a defined order, we have many more ways to specify what we'd like to add or remove, and where in the list we'd like it to go.

One of the simplest operations we can do is to append an element to the end of a list. This looks just like adding an element to a set, correcting for the different literal syntax:

```
UPDATE "user_status_updates"
SET "shared_by" = "shared_by" + ['carol']
WHERE "username" = 'alice'
AND "id" = 76e7a4d0-e796-11e3-90ce-5f98e903bf02;
```

This append operation specifically instructs Cassandra to put carol at the end of the list. If we want to add a value to the beginning of the list instead, we simply flip the order of the values on the right side of the assignment:

```
UPDATE "user_status_updates"
SET "shared_by" = ['dave'] + "shared_by"
WHERE "username" = 'alice'
AND "id" = 76e7a4d0-e796-11e3-90ce-5f98e903bf02;
```

Let's check the current contents of the list to see the effects of our operations:

```
 username | id                                   | shared_by
----------+--------------------------------------+--------------------------
    alice | 76e7a4d0-e796-11e3-90ce-5f98e903bf02 | ['dave', 'bob', 'carol']

(1 rows)
```

The order of the values of the list is exactly what we asked for: dave at the beginning; carol at the end, and bob, the initial sole occupant of the list, taking up the middle.

Writing data at a specific index

What if we want to replace the data at a specific place in the list, rather than adding it to the beginning or the end? We can perform a direct assignment to the index we want as follows:

```
UPDATE "user_status_updates"
SET "shared_by"[1] = 'robert'
WHERE "username" = 'alice'
AND "id" = 76e7a4d0-e796-11e3-90ce-5f98e903bf02;
```

Like array indexes in most programming languages, the index for list element assignment is zero, meaning that in the preceding query we are setting the second element of the list.

List element assignment must reference an index that is within the list's current size. For instance, if we try to assign the fourth element in our current three-element list:

```
UPDATE "user_status_updates"
SET "shared_by"[3] = 'maurice'
WHERE "username" = 'alice'
AND "id" = 76e7a4d0-e796-11e3-90ce-5f98e903bf02;
```

We'll get an out of bounds error:

> **Bad Request: List index 3 out of bound, list has size 3**

> Replacing the value at a given index requires a scan of the list's contents internally, unlike prepending and appending that can be performed without reading any data. This scan makes replacement more expensive than prepending or appending. However, since the actual write is still performed discretely, the operation is still resilient to concurrent updates.

Removing elements from the list

Elements can be removed from a list either by value or by index. Removal by value is identical to set removal, other than the literal format:

```
UPDATE "user_status_updates"
SET "shared_by" = "shared_by" - ['carol']
WHERE "username" = 'alice'
AND "id" = 76e7a4d0-e796-11e3-90ce-5f98e903bf02;
```

This will remove all instances of the string `'carol'` from the list. To remove an element from the list by index, we use a special form of the DELETE statement:

```
DELETE "shared_by"[0]
FROM "user_status_updates"
WHERE "username" = 'alice'
AND "id" = 76e7a4d0-e796-11e3-90ce-5f98e903bf02;
```

Recall from the previous chapter that we can remove data from individual columns in a row by specifying the column names directly after the DELETE keyword; in similar fashion, we can remove data from individual indexes in a list.

Let's check the contents of the list to see the effects of our deletions:

```
username | id                                   | shared_by
---------+--------------------------------------+-----------
   alice | 76e7a4d0-e796-11e3-90ce-5f98e903bf02 | ['robert']

(1 rows)
```

What we've got left is a one-element list. The first deletion, by value, removed `carol` as the last element; the second deletion, by index, removed `david` as the first element. Note that, in both cases, the element in question was completely removed from the list, making the list one element shorter.

 Like writing an element at a specific index, both approaches to remove data from a list require a scan of the list's contents before updating them. Removal of elements from lists is thus slower than other collection manipulation operations, and should be avoided in performance-sensitive code paths.

Using maps to store key-value pairs

The third and final collection type offered by Cassandra is the map, which stores key-value pairs. Keys are unique and unordered, like the elements of a set.

Suppose we'd like to keep track of our users' identities on other social networks. We could create a column for each network, but that would require changing the schema each time we discovered a new network that we want to track, and would also potentially result in a large number of columns in the `users` table just to keep track of a given user's identities.

Instead, let's create a map column that maps the name of a social network to the numeric ID of the user on that network:

```
ALTER TABLE "users"
ADD social_identities MAP<text,bigint>;
```

This definition looks a bit different from the earlier ones in this chapter, since we're now specifying a pair of types rather than just one. Maps do not require their keys and values to be of the same type, so each is specified individually. The first type given is the key type, and the second is the value type. So our keys—the names of social networks—are strings, and our values—IDs on those networks—are long integers.

Writing a map

Like sets and lists, maps have a literal syntax that allows you to directly set the entire contents of a column:

```
UPDATE "users"
SET "social_identities" = {'twitter': 353637}
WHERE "username" = 'alice';
```

The literal syntax for a map separates keys and values with a colon and optional whitespace, delimits key-value pairs with commas, and encloses the whole thing in curly braces. Both keys and values themselves are written in the literal format for their type.

Updating discrete values in a map

When we allow a user to link up to a new social network, we'd like to be able to add their ID to the social_identities collection without regard to its current contents. To do so, we use a syntax similar to that of element replacement in a list:

```
UPDATE "users"
SET "social_identities"['instagram'] = 9839025,
    "social_identities"['yo'] = 25
WHERE "username" = 'alice';
```

It's also perfectly valid to overwrite a value at an existing key; since keys are unique, the previous value will be replaced with the given one:

```
UPDATE "users"
SET "social_identities"['twitter'] = 2725634
WHERE "username" = 'alice';
```

Checking the value of the map in alice's row, we can confirm that the Twitter ID is the most recently written one, along with the Instagram and Yo values:

```
 username | social_identities
----------+---------------------------------------------------------
    alice | {'instagram': 9839025, 'twitter': 2725634, 'yo': 25}

(1 rows)
```

Note that, though the syntax for a map update looks like list replacement, updating a key-value pair in a map does not carry the performance penalty that replacing a list item does.

Removing values from maps

Continuing the parallels, we can remove a value from a map using the DELETE statement, in a similar way to removing items from a list by index:

```
DELETE "social_identities"['instagram']
FROM "users"
WHERE "username" = 'alice';
```

On checking the table again, we can see that the Instagram key-value pair has been removed.

```
 username | social_identities
----------+-------------------------------
    alice | {'twitter': 2725634, 'yo': 25}

(1 rows)
```

> For further reference on the uses of CQL collections, refer to the DataStax CQL documentation at: http://www.datastax.com/documentation/cql/3.1/cql/cql_using/use_collections_c.html.
>
> Collection manipulation operations are documented with the UPDATE statement at http://www.datastax.com/documentation/cql/3.1/cql/cql_reference/update_r.html and the DELETE statement at http://www.datastax.com/documentation/cql/3.1/cql/cql_reference/delete_r.html.
>
> Further information can be found in the official Apache CQL documentation at http://cassandra.apache.org/doc/cql3/CQL.html#collections.

Collections in inserts

So far, all the examples we saw of collection manipulation have used the UPDATE and DELETE statements. It is possible to write the full value of a collection in an INSERT statement as well, by simply providing a literal representation of the collection's contents:

```
INSERT INTO "users" (
  "username", "email", "encrypted_password",
  "social_identities", "version"
) VALUES (
  'ivan',
  'ivan@gmail.com',
```

```
    0x48acb738ece5780f37b626a0cb64928b,
    {'twitter': 875958, 'instagram': 109550},
    NOW()
);
```

The INSERT statement does not, however, support discrete updates. As we explored in the previous chapter, INSERT and UPDATE are, at their core, different ways of expressing the same operation. However, in the context of collection manipulation, as with partial writes, the two statements have different capabilities.

Collections and secondary indexes

Let's say we'd like to get a list of all of the status updates that alice has starred; we can display this as part of her user profile. In order to do this, we'd like to be able to look up all the rows in the user_status_updates table where the starred_by_users column contains the value alice. This is similar to the use case for a secondary index that we explored in the *Using secondary indexes to avoid denormalization* section of *Chapter 5, Establishing Relationships*, except that, in this case, we'd like to be able to perform a lookup based on a single value within a collection column.

Happily, it is entirely valid to put a secondary index on a collection column. The syntax for this is identical to putting an index on any other column:

```
CREATE INDEX ON "user_status_updates" ("starred_by_users");
```

So far, so familiar. Now, we'll introduce the CONTAINS operator, which can be used to look up rows by a value in a collection column:

```
SELECT * FROM "user_status_updates"
WHERE "starred_by_users" CONTAINS 'alice';
```

The WHERE...CONTAINS clause will restrict results to those rows whose starred_by_users set contains the exact value alice. In this case, we'll receive a single result:

```
username | id                                   | body                   | shared_by | starred_by_users
---------+--------------------------------------+------------------------+-----------+-----------------
     bob | 97719c50-e797-11e3-90ce-5f98e903bf02 | Eating a tasty sandwich. |      null |         {'alice'}

(1 rows)
```

Creating secondary indexes on collections is a powerful tool for relating data—now we can model many-to-many relationships using a single, normalized data structure. In this case, the many-to-many relationship is between users and status updates: a user can star many status updates, and a status update can be starred by many users.

It's worth re-emphasizing the warning in the section *Limitations of secondary indexes* of *Chapter 5, Establishing Relationships*. In particular, lookup by a secondary index on a collection column has the same performance characteristics as lookup by a secondary index on a scalar column. As always, it's best to avoid using secondary index lookup for the core access patterns in our applications. In this case, our secondary index enables us to answer the question, "Which status updates has a given user starred?". Since this is not a core feature of the MyStatus application, it's a good use case for secondary indexes.

Secondary indexes on map columns

We just created a secondary index on a set column, which allows us to retrieve rows that contain a given value in that set. Secondary indexes on list columns work in precisely the same way. However, map columns have both keys and values, so it's not immediately obvious what data populates the index.

By default, a secondary index on a map will index the *values* in the map; the CONTAINS keyword similarly filters for map values. However, it is possible to create an index on map keys as well, and then select rows with a given key in a map column.

Let's say we'd like to look up users for whom we have an associated Twitter identity. Recall that in the social_identities column of the users table, the map keys are the names of social networks, and the map values are user IDs from those networks. So, we'll create an index on the keys of the social_identities map:

```
CREATE INDEX ON "users" (KEYS("social_identities"));
```

Note the use of the KEYS operator to indicate that we are creating an index on the map's keys; had we omitted this operator, we would have created an index on the map's values.

Now we can look up all users who have an associated Twitter identity, using the CONTAINS KEY operator:

```
SELECT "username", "social_identities"
FROM users
WHERE "social_identities" CONTAINS KEY 'twitter';
```

The WHERE...CONTAINS KEY clause works exactly the same as WHERE...CONTAINS, except that we are looking up rows that contain the given key in the specified map column. In this case, we get back the two rows with the key twitter in the social_identities map:

```
username | social_identities
---------+-----------------------------------------
    ivan | {'instagram': 109550, 'twitter': 875958}
   alice |             {'twitter': 2725634, 'yo': 25}

(2 rows)
```

The limitations of collections

As we saw, collection columns are a useful and versatile feature of CQL. In fact, we might suppose that collections would be a good way to model data we've modeled in other ways in previous chapters. For instance, why not model a user's status updates as a list of text values in the users table?

As it turns out, collection columns do have some limitations that circumscribe the cases in which they're the best solution. We'll explore the major limitations now.

Reading discrete values from collections

The most powerful feature of CQL collections is the ability to write discrete values to a collection. It's possible to append a single value to a list, update a single key-value pair in a map, remove a specific value from a set, and so on.

At read time, however, there is no special support for reading discrete values. For instance, we might want to be able to do something like the following to read a specific element from a list column:

```
SELECT "shared_by"[2]
FROM "user_status_updates"
WHERE "username" = 'alice'
AND "id" = 76e7a4d0-e796-11e3-90ce-5f98e903bf02;
```

Or we might like to read the value at a specific key within a map column:

```
SELECT "social_identities"['twitter']
FROM "users"
WHERE "username" = 'alice';
```

However, both queries will result in an error; partial reads of collection columns are not possible in CQL. The only way to retrieve data from a collection is to read the collection in its entirety; for this reason, it's generally impractical to store large, unbounded datasets within a collection column. Were we to store status updates in a collection column on the `users` table, we would need to read all of a user's status updates whenever we wanted to display any status update.

Collection size limit

Lack of support for partial reads is one reason to keep collections fairly small; another is that there is a built-in size limit for collections that Cassandra can store. In particular, any given collection can contain no more than 64 KB of data. Nothing will prevent you from inserting more than 64 KB of data into a collection but, when you try to read the collection back, the result will be truncated at 64 KB, resulting in data loss. Of course, a collection that large would be quite unwieldy to work with, so you should stick with collections that are far smaller than that threshold.

In general, data that can be expected to grow in an unbounded fashion is inappropriate for collection columns; stick with datasets that you expect to stay fairly small.

Reading a collection column from multiple rows

Suppose we'd like to retrieve a handful of status updates from the database and display information about them, including who starred them. Naïvely, we might expect to be able to perform a query of this form:

```
SELECT * FROM "user_status_updates"
WHERE "username" = 'alice'
AND "id" IN (
  3f9b5f00-e8f7-11e3-9211-5f98e903bf02,
  3f9b5f00-e8f7-11e3-9211-5f98e903bf02
);
```

However, if we perform this query, we'll see an error:

```
Bad Request: Cannot restrict PRIMARY KEY part id by IN relation as a collection is selected by the query
```

One limitation of Cassandra collections is that you cannot read collections when selecting multiple rows using a WHERE...IN clause. Once you have collection columns in your table, you must explicitly select only noncollection columns when using WHERE...IN.

On the other hand, it's entirely legal to read collection columns when retrieving multiple rows via a range of clustering columns:

```
SELECT * FROM "user_status_updates"
WHERE "username" = 'alice'
ORDER BY "id" ASC
LIMIT 2;
```

This query completes successfully and displays the data from the collection columns:

```
 username | id                                   | body              | shared_by   | starred_by_users
----------+--------------------------------------+-------------------+-------------+------------------
    alice | 76e7a4d0-e796-11e3-90ce-5f98e903bf02 | Learning Cassandra! | ['robert'] | {'bob', 'carol'}
    alice | 3f9b5f00-e8f7-11e3-9211-5f98e903bf02 |      Alice Update 1 |       null |             null

(2 rows)
```

Performance of collection operations

Most collection operations do not involve Cassandra reading data internally; they are purely write-without-reading, which we recall is the most efficient pattern for Cassandra data manipulation. Lists, however, are a partial exception to this rule, in three cases as follows:

- Writing an element at a specific index
- Removing an element at a specific index
- Removing all occurrences of a given value

These three operations do require Cassandra to read the full list before manipulating it, and are therefore slower than other collection manipulations. It's best to avoid these three operations in performance-sensitive code.

All other collection updates are entirely write-without-reading, and can thus be performed very efficiently. In *Appendix A*, *Peeking Under the Hood*, we will take a closer look at exactly how collections are stored and updated.

Working with tuples

We saw that collections are a versatile tool for working with multiple values in a single column. One limitation of collections, however, is that all elements of a collection must be of the same type. If we want to store values of different types in the same collection, we can instead turn to **tuples**.

Unlike collections, which can store an arbitrary number of elements, a tuple column stores a fixed number of elements, each of which has a predefined type. So, while we gain the flexibility to store elements of different types in the same column, we give up flexibility in the number of elements stored.

Creating a tuple column

Let's say we'd like to allow users to add some information about their education to their profile. In particular, users should be able to enter where they went to university, and what year they graduated.

There are two distinct pieces of information here, namely school name and graduation year. However, the two are really two components of a single piece of information. It would make no sense, for instance, to store a graduation year but *not* a school name.

For this reason, education is a good candidate for using a tuple type. The syntax for creating a tuple column is similar to that for creating a collection column: we specify both the fact that we would like to use a tuple type, and the individual types of each component of the tuple:

```
ALTER TABLE "users"
ADD "education" frozen <tuple<text, int>>;
```

This statement introduces two new CQL keywords, `frozen` and `tuple`. Working from the inside out, the `tuple` keyword takes one or more type arguments, separated by commas; each type is applied to a component of the tuple. Order matters here: our `education` tuple has two components, the first of which contains `text`, and the second of which contains an `int`.

The `frozen` keyword means that the `education` column will be stored as a single, indivisible unit. Practically, this means that we cannot insert, change, or remove data in an individual component of the tuple; whenever we write to this column, we write all components together. This stands in contrast to the discrete element modifications we can perform on collection columns.

 In Cassandra 2.1, tuple columns *must* be declared `frozen`. Support for nonfrozen tuple columns that would allow discrete modification of individual tuple components is planned for Cassandra 3.0.

A tuple type declaration, like `frozen <tuple<text, int>>`, can be used anywhere a primitive type like `ascii` or `uuid` is valid. You can create collections of tuples, nest tuples within tuples, or even use tuples as a primary key column.

 For more on tuples, refer to the DataStax CQL documentation:
`http://www.datastax.com/documentation/cql/3.1/`
`cql/cql_reference/tupleType.html`

Writing to tuples

To write data to a tuple column, we provide values for all of the tuple components, in order, separated by commas and grouped by parentheses:

```
UPDATE "users"
SET "education" = ('Big Data University', 2003)
WHERE "username" = 'alice';
```

When we read `alice`'s user record back, `cqlsh` displays tuple data using the same format:

```
 username | education
----------+------------------------------
    alice | ('Big Data University', 2003)

(1 rows)
```

If we wish to populate some of the components of a tuple but leave others empty, we can use the `null` keyword:

```
UPDATE "users"
SET "education" = ('Cassandra College', null)
WHERE "username" = 'bob';
```

As a shorthand, we can omit the last value or values when writing a tuple, as long as we specify at least the first value:

```
UPDATE "users"
SET "education" = ('BDU')
WHERE "username" = 'alice';
```

Note, however, that any components that we omit will be assigned `null` values. Let's take a look at `alice` and `bob`'s rows now:

```
username | education
---------+----------------------------
   alice |                 ('BDU', None)
     bob | ('Cassandra College', None)

(2 rows)
```

Even though we did not explicitly provide a value for the second component of `alice`'s education, the previous value of `2003` was overwritten with a `null` value by our last update. This observation serves to emphasize the `frozen` nature of tuples: any time we update any part of a tuple, we update the entire thing.

Before we move on, let's restore `alice`'s education column to its previous state:

```
UPDATE "users"
SET "education" = ('Big Data University', 2003)
WHERE "username" = 'alice';
```

Indexing tuples

In the previous section, we observed that it would not make much sense for a user to have a graduation year without also naming an educational institution; this provided a rather abstract motivation for grouping those two pieces of information together in a single column. A more practical motivation is the fact that, like any other column, tuples can be indexed.

Let's put an index on our education column; the syntax is no different from creating an index on any other column:

```
CREATE INDEX ON "users" ("education");
```

Armed with this index, we can now perform efficient lookup of user records based on their educational institution and year of graduation:

```
SELECT "username", "education" FROM users
WHERE "education" = ('Big Data University', 2003);
```

As we would hope, the query produces `alice`'s record:

```
username | education
---------+-----------------------------
   alice | ('Big Data University', 2003)

(1 rows)
```

The ability to index tuples provides an important workaround to a constraint we discussed in the section *Limitations of secondary indexes* of *Chapter 5, Establishing Relationships*. As you learned in that chapter, a secondary index can only be applied to a single column. If, for instance, we had separate `school_name` and `graduation_year` columns, it would be impossible to create an index that allowed us to efficiently look up records with a given combination of values in these two columns. Using a tuple, we can place both values in a single column and index that column, giving us much the same effect as a multicolumn index.

User-defined types

Using a tuple column to group educational information into a single column has several advantages, but there's also something missing. The `education` column comprises two components: the first, a string, is the name of an educational institution, and the second, an integer, is a graduation year. As the developers of the MyStatus application, we know what the two components of the tuple represent; however, this knowledge is not made explicit anywhere in the database definition.

To solve this problem, we can instead use a **user-defined type**. User-defined types are very similar to tuples, except that each component has a name. This makes it easier for application developers to easily infer the intent of a user-defined type. As we'll see shortly, user-defined types offer a couple of other advantages over tuples.

Creating a user-defined type

Whereas tuples are specified ad hoc when declaring the type for a column, user-defined types are defined separately and then referred to by name in column definitions. Let's make a user-defined type that we can use for our education column:

```
CREATE TYPE "education_information" (
  "school_name" text,
  "graduation_year" int
);
```

Just like the tuple we declared for our education column, this user-defined type contains two fields: a string containing the name of an educational institution, and an integer containing a year of graduation. Unlike the tuple, the type is declared independently of any particular table or column; it is a first-class entity within the my_status keyspace. This is advantageous when we want to reuse a user-defined type across multiple columns in our keyspace.

 For reference on user-defined types, refer to the DataStax CQL documentation: http://www.datastax.com/ documentation/cql/3.1/cql/cql_reference/ cqlRefUDType.html

Assigning a user-defined type to a column

Since the education_information type is designed to play the same role as the tuple used in our existing education column, let's drop the tuple column and replace it with the more semantic user-defined type:

```
ALTER TABLE "users" DROP "education";


ALTER TABLE "users"
ADD "education" frozen <"education_information">;
```

Observe that, as with tuples, we must use the frozen keyword when creating columns with user-defined types. As earlier, this makes explicit the fact that an education_information value is unitary, despite having an internal structure comprising multiple fields.

Adding data to a user-defined column

The syntax for a user-defined type literal is similar to that of a map literal, except that the keys are identifiers rather than literal values. For instance, we can re-add alice's education information into our new column as follows:

```
UPDATE "users"
SET "education" = {
  "school_name": 'Big Data University',
  "graduation_year": 2003
}
WHERE "username" = 'alice';
```

Now we can observe that the education information is present in alice's profile:

```
username | education
----------+-------------------------------------------------------------
   alice | {school_name: 'Big Data University', graduation_year: 2003}

(1 rows)
```

Indexing and querying user-defined types

Like tuple columns, columns with user-defined types can be indexed and queried. Let's add an index to our new education column:

```
CREATE INDEX ON "users" ("education");
```

Now we can query for a given value of education keeping in mind that, just as with a tuple, we must specify the full value of the user-defined type, with all fields:

```
SELECT "username", "education" FROM "users"
WHERE "education" = {
  "school_name": 'Big Data University',
  "graduation_year": 2003
};
```

As expected, the result consists of alice's row, which contains the exact value we specified in her education column:

```
username | education
----------+-------------------------------------------------------------
   alice | {school_name: 'Big Data University', graduation_year: 2003}

(1 rows)
```

Partial selection of user-defined types

In some situations, we might only be interested in what school alice went to, but not her year of graduation. In this case, we would like to be able to perform a query that returns precisely the information we're interested in, without wasting network bandwidth on irrelevant data. Happily, user-defined types have this capability.

In order to do this, we simply use a dot operator to specify a single field within a user-defined type that we would like to return. For instance, to retrieve only the `school_name` field from `alice`'s `education` field, we can perform the following query:

```
SELECT "username", "education"."school_name"
FROM "users"
WHERE "username" = 'alice';
```

Now, the school name is presented as its own column in the results:

```
username | education.school_name
----------+-----------------------
   alice |    Big Data University

(1 rows)
```

Note that, internally, a user-defined type is stored as a single, indivisible entity; in order to return a partial value for a user-defined type, Cassandra must, under the hood, deserialize the entire value and then extract the field we requested. The advantage here is merely that of avoiding the network bandwidth of transferring extraneous fields within the user-defined type.

Choosing between tuples and user-defined types

As we saw, tuples and user-defined types have a lot in common. In both structures, a column contains a fixed, predefined set of fields, each of which can have its own type. Both structures are stored as `frozen`, meaning that Cassandra cannot perform discrete operations on their internal components. Both can be indexed and used in the WHERE clause of a query. So how do we decide which to use?

In most cases, a user-defined type is a better option. User-defined types give names to their fields, making it easier for application developers to reason about their usage. Also, user-defined types can be partially selected in queries; tuples cannot.

The only reason to use a tuple is convenience: a tuple does not need to be defined separately from its use in a column definition. So, for quick prototyping of schema structures, a tuple can be a better option. However, for a schema that's going into production, a user-defined type is nearly always going to be the right choice.

Comparing data structures

We've now seen five different data structures that allow us to store multiple values within a single field. Each data structure has certain advantages and disadvantages, and each is best suited to particular use cases. In the table below, we summarize the important features of each of the data structures from this chapter:

	Set	List	Map	Tuple	User-defined Type
Size	Flexible	Flexible	Flexible	Fixed	Fixed
Discrete updates	Yes	Yes	Yes	No	No
Partial Selection	No	No	No	No	Yes
Name-value pairs	No	No	Yes	No	Yes
Multiple Types	No	No	Keys and Values	Yes	Yes
Index	Individual elements	Individual elements	Individual elements	Entire value	Entire value
Can be used as primary key	No	No	No	Yes	Yes

Summary

Collection columns are a powerful feature of CQL that allow us to store multiple values in a single column. Most importantly, it's possible to discretely update single values in a collection without reading the collection's current contents or fully providing the new contents of the collection.

This capability is particularly useful when multiple processes might need to concurrently modify a collection. By avoiding the need to read and then fully overwrite a collection's contents, we avoid situations in which concurrent updates can lead to data loss and can support concurrent updates without resorting to optimistic locking.

Collections are best suited to datasets that are small and bounded. This is both because there is a hard upper limit on the amount of data a collection can hold, and because, when a collection is read, it is always read in full. For larger data sets, it is usually most appropriate to create a separate table whose partition key reflects the full primary key of the parent row, as we covered in *Chapter 3, Organizing Related Data*.

Tuples and user-defined types are both data structures that store a fixed number of values in a column. Values can be of different types, but they must be of the exact number and types declared in the column definition. Tuples and user-defined types are particularly useful in the way that they can be indexed, providing a workaround for the rule that a secondary index can only apply to a single column. In most situations, user-defined types are a better choice than tuples because they offer the additional benefit of named fields and partial selection.

In the next chapter, we will explore another special type of column that allows us to increment and decrement counter information without reading first. Like collection columns, counter columns are useful primarily because we can perform discrete mutations on their values without knowing what the value is. Counter columns are particularly useful for time-series statistical aggregation; we'll explore data structures for this use case next.

9

Aggregating Time-Series Data

In the preceding chapters, you learned how to use Cassandra as a primary data store for the MyStatus application, with a focus on modeling the data that drives the main user experience on the site. In this chapter, we'll shift focus to another popular use of Cassandra: aggregating data that we observe over time. In particular, we'll build a small analytics component into our schema, allowing us to keep track of how many times a given status update was viewed on a given day.

In order to do this, we'll introduce a new type of column, the counter column, which is a special numeric column type that can be discretely incremented or decremented. Counter columns have a lot in common with collection columns, which we explored in the previous chapter: you can make discrete changes to them without reading their current value, and they're good for scenarios in which many threads or processes might need to update the same piece of data at the same time.

There are also big differences between counter columns and collection columns: most obviously, counter columns only hold a single scalar value, whereas collection columns hold multiple values. There are also more subtle differences and limitations that we'll explore in this chapter.

By the time you're done with this chapter, you'll have learned:

- How to construct tables for time-series aggregation
- How to define a counter column
- How to increment and decrement a counter column
- Limitations of counter columns

Recording discrete analytics observations

Let's say we want to keep very close track of how many times our users' status updates are viewed by someone else. Status updates may be viewed on the MyStatus web site, or by using our mobile app, or via a third-party app using our API. We'll want to track that, as well as which status update was viewed and when. To do this, let's create a table to store analytics observations:

```
CREATE TABLE "status_update_views" (
  "status_update_username" text,
  "status_update_id" timeuuid,
  "observed_at" timeuuid,
  "client_type" text,
  PRIMARY KEY (
    ("status_update_username", "status_update_id"),
    "observed_at"
  )
);
```

In this new table, we store a partition for each individual status update, with the full primary key of the status update serving as the partition key for our table. Each time someone views a status update, we'll store a new row in the table, generating a timestamp UUID for the row to populate the observed_at clustering column.

We'll also, for each observation, keep track of what type of client was used to view the status update: the web site, our mobile app, or a third-party API client. We could, of course, keep track of many other bits of metadata about the observation, such as the IP address of the user, the exact browser or app version being used, and so on.

Using discrete analytics observations

The status_update_views table gives us a complete and highly discrete view of the usage data we're collecting but it is limited in what questions it can answer. Of course, by analyzing the primary key structure, we know that the question it is best suited to answer is: "What do we know about each view of a given status update in a given time range?" We can answer this question using a range slice query of the following form:

```
SELECT "observed_at", "client_type"
FROM "status_update_views"
WHERE "status_update_username" = 'alice'
  AND "status_update_id" = 76e7a4d0-e796-11e3-90ce-5f98e903bf02
  AND "observed_at" >= MINTIMEUUID('2014-10-05 00:00:00+0000')
  AND "observed_at"  < MINTIMEUUID('2014-10-06 00:00:00+0000');
```

This query will give us information about all the observed status update views on October 5, 2014. We know that, because we are asking for rows in a single partition and a specified range of clustering columns, Cassandra can perform the query very efficiently.

What about other questions we'd like to ask regarding the user behavior we've observed? One thing we can do is simply ask how many total status updates have been observed in a given time range:

```
SELECT COUNT(1)
FROM "status_update_views"
WHERE "status_update_username" = 'alice'
  AND "status_update_id" = 76e7a4d0-e796-11e3-90ce-5f98e903bf02
  AND "observed_at" >= MINTIMEUUID('2014-10-05 00:00:00+0000')
  AND "observed_at"  < MINTIMEUUID('2014-10-06 00:00:00+0000');
```

Compared with retrieving the data in each row and then counting the number of rows we get back, we do save some data transfer using a COUNT query. However, under the hood, Cassandra is doing roughly the same thing: reading all the rows in the table that match our criteria and then counting them.

What's worse, since CQL does not offer anything like the GROUP BY clause in SQL, is that we can only get *one* count back at a time. If we want to know how many views a given status update has received on each day in September, we either need to perform thirty COUNT queries, or we need to read all the observation rows for September and then calculate aggregate counts in our application.

Slicing and dicing our data

There are many interesting questions we can ask about our analytics observations; most of them can't efficiently be answered by the status_update_views table. Here are a few:

- How many total views did status updates receive on each day of September?
- What percentage of status update views are via the web, via our mobile app, and via third-party applications?
- Which hours of the day are most popular for viewing status updates?
- What is the average monthly growth of status update views this year?

In order to answer any of these questions efficiently, we will require a precomputed table that's structured with that question in mind. Instead of reading massive quantities of raw observations into memory, and then computing aggregate information about it, we'll keep the aggregate information up-to-date as we make the discrete observations.

Recording aggregate analytics observations

When we explore precomputed aggregate data storage, let's focus on the first question posed above: how many total views did status updates receive on each day of September?

More generally, we want to store the daily overall view counts in a structure that allows us to easily retrieve the counts for a given range of time. We don't need to store discrete information about every view event that happened; simply knowing how many views occurred per day is sufficient.

Let's create a new `daily_status_update_views` table that aggregates our analytics observations at just the right granularity:

```
CREATE TABLE "daily_status_update_views" (
  "year" int,
  "date" timestamp,
  "total_views" counter,
  "web_views" counter,
  "mobile_views" counter,
  "api_views" counter,
  PRIMARY KEY (("year"), "date")
);
```

Of course, the most striking thing about this table definition is the introduction of the `counter` column type; we'll dive into this a little later in this chapter. First, let's take a close look at the structure of the primary key and the data columns.

Answering the right question

Recall that the job of this table is to be able to answer questions such as, "How many total status update views happened on each day of September?". In order to do so, we need to store view counts aggregated daily, and we need to be able to query for ranges of days.

By forming a primary key out of the year, month, and day on which the views were counted, we aggregate at the right granularity: there will be one row per day, containing the counts for that day. We use the year of the observation as the partition key, thus giving ourselves the ability to retrieve a series of daily counts for any range of days within the same year. In the event that we want to retrieve a series of daily counts for a range that spans multiple years, our application will need to make multiple queries—one for each year we're interested in— and then stitch together the results on the client side.

We use a timestamp column as the clustering column, storing the day we're counting views on. The information in the `year` partition key is redundant with the year component of the clustering column, so we'll never need to read it directly, but year-based partitions still allow us to keep the size of each partition reasonable and bounded.

For instance, if we want to retrieve total view counts for each day over the month of September, it's as simple as this:

```
SELECT "date", "total_views"
FROM "daily_status_update_views"
WHERE "year" = 2014
  AND "date" >= '2014-09-01'
  AND "date"  < '2014-09-30';
```

Another fact about the schema that's worth noting is the use of four data columns: a total count, and three columns for client-specific counts. Another way to model the same information would have been to add another clustering column called, say, `client_type`; then, for each day, we'd have four rows—one per client type. One approach is not obviously better than the other; in fact, as we'll see in the *Appendix A, Peeking Under the Hood*, the two schemas are more similar than we might imagine.

An advantage of the approach we took is that there's exactly one row per day, which is easier to reason about and more naturally fits our intuitions about the data. An advantage of the one-row-per-client-type approach is that we can introduce new client types without having to add additional columns to the schema. If we expect that new client types will be introduced regularly, storing the client type in a clustering column, rather than as part of the schema definition, becomes very appealing.

Precomputation versus read-time aggregation

Now we have a table that's optimized to aggregate our analytics data in a specific way. The advantage of this approach is that we can now answer questions about daily views very efficiently; the downside is that we need to create and maintain a table just to aggregate the data in this one way. Each time we record a view, we need to update both the `status_update_views` and `daily_status_update_views` tables. We can easily imagine dozens of different ways in which we might want to aggregate analytics data, each requiring its own purpose-built table, each needing to be updated when an observation is made.

Cassandra is well suited to this sort of **precomputed aggregation** because of properties that we have explored in previous chapters. As we can horizontally scale our dataset by simply adding more machines to our cluster, storing many different aggregates of the same underlying observations isn't hugely expensive. As Cassandra is extremely efficient at writing data, it isn't a deal-breaker to have to update several different tables when we make an observation. The tradeoff is analogous to the one we considered in our exploration of data denormalization in *Chapter 6, Denormalizing Data for Maximum Performance*: by keeping multiple views of the same underlying data, we increase the complexity of writing data, but give ourselves very efficient structures from which to read that data.

An alternative to precomputation is aggregating data at read time. SQL databases give us a substantial toolset for performing ad hoc aggregation of data when reading it back, using aggregate functions like SUM, AVERAGE, MININUM, MAXIMUM, and so on, combined with a GROUP BY clause to perform the aggregations at the desired level of granularity. CQL does not offer built-in aggregation functionality; however, Cassandra is capable of integrating with the Hadoop MapReduce framework, which provides an efficient means of performing aggregate computations over massive datasets. DataStax Enterprise, a commercial package that bundles Cassandra, Hadoop, and the Solr search engine, is worth exploring if you need to aggregate data in an ad hoc way.

The many possibilities for aggregation

Having satisfied ourselves that precomputed aggregation is worth adding some complexity to the process of recording an observation, let's explore some other ways in which we might want to aggregate data. Here are a few examples of questions we might want to answer, and the type of table schema we might set up to efficiently store precomputed aggregates to answer these questions:

Question	Partition Key	Clustering Column
How many total views did status updates receive in September?	year	date
What percentage of status update views are via the web, via our mobile app, and via third-party applications?	month	client_type
Which hours of the day are most popular for viewing status updates?	month	hour_of_day
What is the average monthly growth of status update views this year?	year	month

In each case, we choose a partition key that divides our observations into large chunks of time that we'll most likely want to query within, rather than across; then we choose a clustering column that aggregates data using the appropriate level of granularity.

The role of discrete observations

One question that might come to concern us is the purpose of the first table we created in this chapter: `status_update_views`. As we observed, this table isn't terribly useful for answering any aggregate questions about our usage data, so why store it at all?

As it turns out, we probably won't interact directly with `status_update_views` when exploring our analytics observations on a day-to-day basis. However, `status_update_views` stores the raw material for all of the aggregate tables. If, in a few months, we decide that we'd like to aggregate data on a previously unforeseen dimension, we can backfill our aggregates using the raw observations in `status_update_views`. Keeping the raw data around gives us some flexibility when it comes to designing new aggregates down the road.

Recording analytics observations

At this point, we've explored structuring both discrete and aggregate analytics data, and looked at accessing that data. However, to have interesting data aggregates to access, we first need to record our observations.

Updating a counter column

The `daily_status_update_views` table introduces a new type of column: the counter column. Counter columns store integer values, just like `int` and `bigint` columns; however, unlike a normal data column, counter columns are always incremented or decremented, rather than having a value set directly.

We've currently got two tables to store usage data: `status_update_views` to store raw view observations, and `daily_status_update_views` to store views by day. We'd like to record that one of `alice`'s status updates was viewed on the web on October 5, 2014 at 3:12 P.M. EDT:

```
INSERT INTO "status_update_views" (
  "status_update_username", "status_update_id",
  "observed_at", "client_type"
) VALUES (
  'alice', 76e7a4d0-e796-11e3-90ce-5f98e903bf02,
  85a53d10-4cc3-11e4-a7ff-5f98e903bf02,
```

```
        'web'
);

UPDATE "daily_status_update_views"
SET "total_views" = "total_views" + 1,
    "web_views" = "web_views" + 1
WHERE "year" = 2014
  AND "date" = '2014-10-05';
```

The first INSERT statement should look familiar: we're adding a new row whose partition key columns contain the full primary key of the status update that was viewed, and whose clustering column is a timestamp UUID encoding the exact time at which the status update was viewed. We also record what kind of client was used to view the status update; in this case, a web browser.

The second statement, UPDATE, introduces the manipulation of counter columns. Syntactically, we've seen this before; specifically, when adding an item to a list or set column, we used the same syntax of *column_name = column_name + incremental_value*. In this case, however, the incremental value is an integer, rather than one or more items we want to add.

The overall effect is the same: the values of the total_views and web_views columns are incremented by one each. As with discrete modification of collection columns, at no point did we need to read the current value of the column into memory; for this reason, we don't need to worry about the sort of race condition we explored in the previous chapter.

Let's take a look at the contents of the daily_status_update_views table:

```
SELECT * FROM "daily_status_update_views";
```

As we can see, our observation has been recorded for the day in question:

year	date	api_views	mobile_views	total_views	web_views
2014	2014-10-05 00:00:00-0400	null	null	1	1

(1 rows)

You'll notice that the date column contains not just a date, but also a time. CQL does not offer a built-in date type, but does allow us to specify timestamp literals using only a date; it will interpret this as midnight on that day in the system time zone of the machine on which Cassandra is running.

Counters and upserts

We'll also note that our earlier UPDATE statement attempted to increment a column in a row that did not yet exist. As we saw in *Chapter 7, Expanding Your Data Model*, this is an upsert operation; the UPDATE statement actually creates the row and increments the values accordingly. Not surprisingly, if we attempt to increment a counter column that doesn't yet have a value, its initial value is considered zero.

Setting and resetting counter columns

In certain situations, we might want to set a counter column directly; for instance, if we are backfilling our daily_status_update_views table from historical information in status_update_views, one approach would be to calculate the view counts for each day in our application, and then write the counts directly to daily_status_update_views.

The natural way to do this would be to issue an INSERT statement to put the known total view count in the row for a given day:

```
INSERT INTO "daily_status_update_views"
("year", "date", "total_views")
VALUES (2014, '2014-02-01', 500);
```

However, if we try to perform this operation, we'll see an error:

```
Bad Request: INSERT statement are not allowed on counter tables, use UPDATE instead
```

As it turns out, the *only* way to mutate a counter column value is to increment or decrement it; the value cannot be set directly. For the backfill use case, our best bet is just to issue an increment statement for each of the raw views we have stored. If we know that the daily_status_update_views table has no information stored for the day in question prior to the backfill, we can also simply perform a single increment, since adding our computed aggregate to the current value of zero will effectively insert the computed aggregate:

```
UPDATE "daily_status_update_views"
SET "total_views" = "total_views" + 500
WHERE "year" = 2014
  AND "date" = '2014-02-01';
```

On checking our daily aggregates table again, we'll see our upsert has had the intended effect:

```
year | date                     | api_views | mobile_views | total_views | web_views
------+--------------------------+-----------+--------------+-------------+----------
2014 | 2014-02-01 00:00:00-0500 |      null |         null |         500 |      null
2014 | 2014-10-05 00:00:00-0400 |      null |         null |           1 |         1
```

(2 rows)

Counter columns and deletion

Just as INSERT statements are not a useful operation on counter columns, neither are DELETE statements. It is possible to delete a row containing counter columns:

```
DELETE FROM "daily_status_update_views"
WHERE "year" = 2014
   AND "date" = '2014-02-01';
```

However, a counter column deletion is permanent: you can never put data back in the counter in that row. Let's try re-adding some data to the row for February 1:

```
UPDATE "daily_status_update_views"
SET "total_views" = "total_views" + 100
WHERE "year" = 2014
   AND "date" = '2014-02-01';
```

On reading the data in the table, we'll see something surprising:

```
year | date                     | api_views | mobile_views | total_views | web_views
------+--------------------------+-----------+--------------+-------------+----------
2014 | 2014-10-05 00:00:00-0400 |      null |         null |           1 |         1
```

(1 rows)

The data we've just inserted isn't there. Cassandra reported no error when we tried to increment the counter; it just silently dropped the operation.

Because of the way counter columns handle deletions, it's inadvisable to ever issue a DELETE statement against a counter column. Counters are great for aggregating values that you will never need to reset or remove; as long as you only need to modify counters via increments and decrements, they perform admirably. Aggregating user behavior is one use case that precisely fits within this domain.

Counter columns need their own table

Another surprising fact about counter columns is that they cannot coexist with regular data columns in the same table. For instance, let's imagine that, along with view counts, we'd like to record the timestamp of the last view of a status update on a given day. We might think we can add a timestamp column to our `daily_status_update_views` table:

```
ALTER TABLE "daily_status_update_views"
ADD "last_view_time" timestamp;
```

However, if we attempt to perform this schema modification, we'll see an error:

```
Bad Request: Cannot add a non counter column (last_view_time) in a counter column family
```

Cassandra requires that counter columns live in their own table; they cannot coexist with regular data columns or collection columns. Of course, counter tables still have primary key columns with the full range of data types; in fact, primary key columns cannot have the counter type, since using a primary key that cannot be directly set would not make sense.

Summary

In this chapter, we explored strategies for aggregating observed time-series data—in this case user behavior in viewing status updates in our application. While user behavior analytics are a fantastic and common use case for Cassandra, we could also take the same approach to aggregate scientific data, economic data, or anything else where we'd like to roll up discrete observations into high-level aggregate values.

Our structure for recording time-series data used a table containing discrete observations as the raw material and acting as the data record in case we want to introduce new aggregate dimensions down the line. We also used a table that precomputed aggregate observations by day; by keeping the aggregate up-to-date at write time, we built a structure that allows us to very efficiently retrieve aggregates over a given time period, without any expensive computation at read time. We can easily imagine constructing dozens of such tables, one for each level of granularity at which we would like to analyze aggregate information.

We explored using counter columns to effortlessly maintain the precomputed aggregates; each time we made an observation, we simply issued an upsert to increment the relevant counter columns; this allowed us to record observations simply by issuing a series of UPDATE statements, without having to read the current aggregate values from Cassandra first.

We saw that, while counter columns are a useful tool for precomputed data aggregation, they also have their downsides. Counter columns do not allow us to directly set values, we can only increment or decrement them; because deletion of a counter column value is permanent, this operation is of little use in a counter column table. We saw that counter columns can coexist in a table only with other counter columns; they can't be in the same table as other data columns or collection columns.

In the next chapter, we will look more deeply into how Cassandra stores and retrieves data, with particular focus on how data is distributed among multiple machines in a multinode cluster, the typical configuration of a production Cassandra deployment. You'll learn how Cassandra handles conflicting updates to the same piece of data using timestamps, and you'll see how we can override those timestamps to interesting effect. You'll also learn more about what happens when data is deleted from Cassandra, and use that knowledge to avoid common pitfalls with data deletion.

10
How Cassandra Distributes Data

Much of Cassandra's power lies in the fact that it is a **distributed database**: rather than storing all of your data on a single machine, it is designed to distribute data across multiple machines. A distributed architecture is hugely beneficial for scalability since you're not bound by the hardware capacity of a single machine; if you need more storage or more processing power, you can simply add more nodes to your Cassandra cluster. It's also a boon for availability: by storing multiple copies of your data on multiple machines, Cassandra is resilient to the failure of a particular node.

The beauty of a distributed database such as Cassandra is that, as application developers, we rarely need to think about the fact that we're working with data that's spread across multiple servers. We've spent the last nine chapters exploring a wide range of Cassandra's functionality, and the interfaces we've worked with never require us to explicitly account for the fact that data is distributed. From the application's perspective, we simply write data to Cassandra and then read it back; the database takes care of figuring out which machine or machines the data is written to or read from.

That said, when developing applications using Cassandra as a persistence layer, it's important to understand how data is distributed and replicated. One topic of keen interest to application developers is **consistency**: if multiple copies of a piece of data exist on different machines in the cluster, how do I know that I'm reading the most up-to-date version of the data? In distributed data stores, there is always a tradeoff between consistency and availability; Cassandra provides tunable consistency, which allows us to decide which is more important in any given scenario.

Another important consideration in a distributed data store is conflict resolution: if two clients attempt to write different data to the same location at the same time, which write wins? Under the hood, every piece of data written to Cassandra has a timestamp attached to it; conflicts are resolved with a simple last-write-wins strategy. Application developers have the ability to override the default timestamp attached to a write operation, sometimes to interesting effect.

Finally, a fault-tolerant distributed database needs to take special care when deleting data. In particular, a deletion does not result in Cassandra immediately forgetting that a value ever existed; doing so might lead to the data unexpectedly reappearing in certain scenarios, which we will explore later in the chapter. Instead, Cassandra creates a **tombstone**, which encodes the knowledge that a certain value was deleted at a certain time. Understanding tombstones is crucial to avoiding certain performance-killing Cassandra anti-patterns.

With a solid understanding of how Cassandra distributes data and a keen eye on how those mechanisms affect our work as application developers, you'll be fully prepared to build highly scalable, efficient, and fault-tolerant applications using Cassandra. In this chapter, you will study:

- How Cassandra distributes data in a multi-node cluster
- How Cassandra replicates data
- The difference between immediate consistency and eventual consistency
- The tradeoff between consistency and availability
- How and when to tune consistency
- How Cassandra resolves conflicting writes
- How to provide your own timestamp when writing data
- How and why Cassandra uses tombstones for data deletion
- Writing expiring data using TTLs

A Word About Examples

In the preceding chapters, we've consistently used hands-on examples in the MyStatus application to demonstrate the features of Cassandra under discussion. In this chapter, we will be covering concepts whose effects are largely transparent to an application under normal circumstances. For that reason, much of the material in this chapter will consider specific scenarios — usually adverse ones — in a multi-node cluster, which can't be easily demonstrated using the single-node development cluster you set up in *Chapter 1, Getting Up and Running with Cassandra*.

The material in this chapter isn't entirely hands-off: you'll get a chance to play with timestamps and TTLs and observe their effects in the cqlsh console. But most of this chapter is devoted to developing a mental model of how your data behaves in a production environment, so you'll need to use your imagination a bit, amply aided by diagrams and illustrations.

Data distribution in Cassandra

In a traditional relational database such as MySQL or PostgreSQL, the entire contents of the database reside on a single machine. At a certain scale, the hardware capacity of the server running the database becomes a constraint: simply migrating to more powerful hardware will lead to diminishing returns.

Let's imagine ourselves in this scenario, where we have an application running on a single-machine database that has reached the limits of its capacity to vertically scale. In that case, we'll want to split the data between multiple machines, a process known as **sharding** or **federation**. Assuming we want to stick with the same underlying tool, we'll end up with multiple database instances, each of which holds a subset of our total data. Crucially, in this scenario, the different database instances have no knowledge of each other; as far as each instance is concerned, it's simply a standalone database containing a standalone dataset.

It's up to our application to manage the distribution of data among our various database instances. Specifically, for any given piece of data, we'll need to know which instance it belongs on; when we're writing data, we need to write to the right instance, and when we're reading data, we need to read it back from that same place.

Assuming we are using integer primary keys—the standard practice in relational databases—one simple strategy is to use the modulus of the primary key. If we have four database instances, our primary key layout would look something like this:

Instance	Primary Keys
Instance 1	1, 5, 9, 13, …
Instance 2	2, 6, 10, 14, …
Instance 3	3, 7, 11, 15, …
Instance 4	4, 8, 12, 16, …

Of course, this strategy would likely result in great difficulty performing relational operations; it's quite likely that a foreign key stored on one instance would refer to a primary key stored on a different instance. For this reason, in a real application we would likely design a more sophisticated partitioning strategy. Regardless of what we choose, the critical feature of any partitioning strategy is that we know which database instance a given piece of data belongs on, being given only the information we have before actually communicating with a database instance.

Cassandra's partitioning strategy: partition key tokens

As application developers working with Cassandra, we never need to go through the above calculus: when we write or read data, we can perform the query on any node in the cluster, and Cassandra will figure out where the data lives. The federation process is entirely transparent to the application.

As it turns out, Cassandra uses a strategy analogous to the naïve primary key modulus approach described above. Recall from *Chapter 2, The First Table,* that Cassandra has a TOKEN function that generates an integer value for any partition key; when we retrieve results over multiple partition keys, the rows are ordered by this token.

Distributing partition tokens

When Cassandra distributes data, it assigns each node a range of tokens; a row is stored on the node within whose token range its partition key token falls. Since tokens are generated using a hashing function, token values are distributed evenly across the entire range of possible values. So, as long as the number of partition keys is much bigger than the number of nodes in the cluster, partition keys will be balanced evenly between the different nodes, if each node is responsible for an equally sized portion of the token range.

Let's take a look at a few rows in our `users` table and examine how they would be distributed in a three-node cluster:

```
SELECT "username", TOKEN("username")
FROM "users"
WHERE "username" IN ('alice', 'bob', 'ivan');
```

We'll see that these three rows have primary key tokens that are pretty evenly distributed across the 64-bit space of possible tokens:

```
 username |  token(username)
----------+--------------------
    alice |  5699955792253506986
      bob | -5396685590450884643
     ivan |   962209788683003613
```

Now let's consider how data will be distributed across our three-node cluster. Cassandra tokens are signed 64-bit integers, so the minimum possible hash is -2^{63} or -9223372036854775808 and the maximum possible hash is $2^{63}-1$ or 9223372036854775807. So, our nodes will be assigned to token ranges as follows:

Node	Lowest Token	Highest Token
Node 1	-9223372036854775808	-3074457345618258604
Node 2	-3074457345618258603	3074457345618258601
Node 3	3074457345618258602	9223372036854775807

It's common to visualize a topology as a ring, with each range in the token space corresponding to a portion of the circumference of the ring, much like degrees in an arc. We can apply that visualization to our three-node cluster like this:

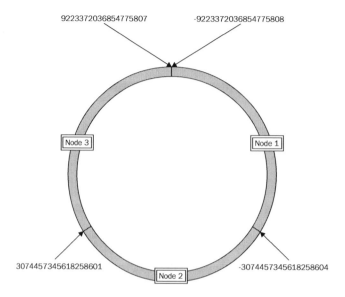

Now let's consider the location of each of the three rows we looked at previously. As it turns out, each row falls in a different range in the token space: alice lives on **Node 1**, ivan lives on **Node 2**, and bob lives on **Node 3**. Using our ring diagram, we can visualize how each row is assigned to a node:

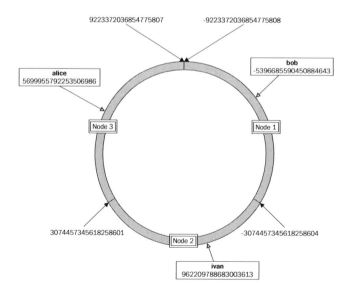

Partition keys group data on the same node

In *Chapter 3, Organizing Related Data,* you learned that tables with compound primary keys store all rows sharing the same partition key in contiguous physical storage. This leads to the observation that querying for ranges of clustering column values within a single partition key is highly efficient. To perform this sort of lookup, Cassandra need only locate the beginning of the range on disk, and can then read all the results beginning at that location. Conversely, querying for rows spanning multiple partition keys requires an inefficient random disk scan for each partition key being queried.

You new understanding of data partitioning expands this observation: you now know that querying for multiple partition keys not only requires Cassandra to make multiple disk scans, but very likely will also require retrieving data from multiple nodes and collating the results. Cassandra is entirely capable of performing this operation—the process of reading from multiple nodes and collating the results is performed by a **coordinator node** and is entirely transparent to the application. But it's important to remember that the process of reading data from multiple partitions—and thus possibly multiple nodes—is expensive and best avoided for performance-sensitive operations.

Virtual nodes

The model of data distribution we have developed thus far is, in fact, a simplification of how a modern Cassandra cluster works. While versions of Cassandra prior to 1.2 did directly map ranges of tokens onto physical nodes, Cassandra 1.2 introduced **virtual nodes**, which act as an intermediary in the mapping process.

Virtual nodes replace physical nodes in the partitioned ring; each virtual node owns a portion of the token space. Virtual nodes themselves are then owned by physical nodes, but a physical node does not own a contiguous range of virtual nodes; rather, virtual nodes are distributed randomly among physical nodes. Crucially, there are many more virtual nodes than physical nodes; each physical machine is responsible for many virtual nodes.

Looking back at our three-node data cluster, let's examine how data is distributed using virtual nodes. For simplicity, we'll say there are twelve virtual nodes in the cluster—four per physical node—although in a real cluster that number would be much higher. The makers of Cassandra recommend 256 virtual nodes in a production cluster.

Here's how our ring now looks, divided up into twelve virtual nodes:

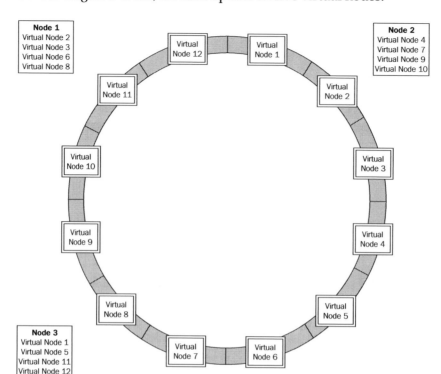

Note that each physical machine is no longer responsible for a contiguous range of tokens; instead, each machine is responsible for four virtual nodes, each of which covers a different token range.

Virtual nodes facilitate redistribution

The main advantage of virtual nodes is their behavior when the cluster changes. Consider the simple example of adding a fourth node to our three-node cluster. Without virtual nodes, all three preexisting physical nodes have their token range changed to make space for the fourth node. To accommodate the fourth machine, Cassandra must recalculate the target node for each individual row based on the mapping from its token to the new token range assignments of the four nodes. This is a process known as **rebalancing**, and it's rendered unnecessary by virtual nodes.

When a new node joins the cluster, it's simply assigned a handful of virtual nodes that previously belonged to other machines. Rather than directly recalculating the physical location of each individual row, Cassandra can simply assign the correct number of virtual nodes to the new machine—in this case, three—and move their contents over wholesale. Unlike in a rebalancing scenario, where every physical machine is both losing and gaining data, redistributing virtual nodes only requires data to be moved from the original three machines to its new home on the fourth machine. Here's how the ring will now look:

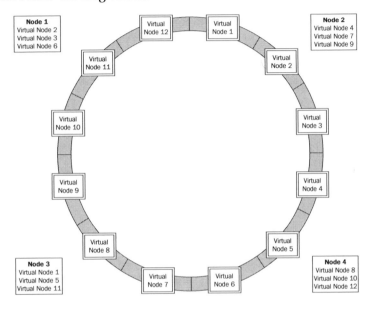

 While a treatment of virtual nodes is important to cultivate a complete understanding of how Cassandra data distribution works, it's worth emphasizing that the process of accommodating changes to cluster topology—such as adding, removing, or replacing nodes—is entirely transparent to the application. Nodes can be added to, or removed from, a live Cassandra cluster with no degradation of functionality from the application's standpoint. The same is true for unexpected changes to the cluster, such as the failure of a node, thanks to Cassandra replication, which we'll cover next.

Data replication in Cassandra

So far, we've developed a model of distribution in which the total data set is distributed among multiple machines, but any given piece of data lives on only one machine. This model carries a big advantage over a single-node configuration, which is that it's horizontally scalable. By distributing data over multiple machines, we can accommodate ever-larger data sets simply by adding more machines to our cluster.

But our current model doesn't solve the problem of fault-tolerance. No hardware is perfect; any production deployment must acknowledge that a machine might fail. Our current model isn't resilient to such failures: for instance, if **Node 1** in our original three-node cluster were to suddenly catch fire, we would lose all the data on that node, including the row containing `alice`'s user record.

To solve this problem, Cassandra provides replication; in fact, no serious Cassandra deployment would store only one copy of a given piece of data. The number of copies of data stored is called the **replication factor**, and it's configured on a per-keyspace level. Recall the query that we used to create our `my_status` keyspace in *Chapter 1, Getting Up and Running with Cassandra*:

```
CREATE KEYSPACE "my_status"
WITH REPLICATION = {
  'class': 'SimpleStrategy',
  'replication_factor': 1
};
```

For our development environment, we chose a replication factor of 1; there is little reason to store multiple copies of the data, since we're only using a single node for development. In a production deployment, however, we would choose a higher number; 3 is a good default.

Masterless replication

If you've worked with a relational database in production, it's likely you have experience with replication. Relational databases typically provide **master-follower replication**, in which all data is written to a single master instance; then, behind the scenes, the writes are replicated to follower instances. The application can read data from any of the followers.

Note that master-follower databases are not distributed: every machine has a full copy of the dataset. Master-follower replication is great for scaling up the processing power available for handling read requests, but does nothing to accommodate arbitrarily large datasets. Master-follower replication also provides some resilience against machine failure: in particular, failure of a machine will not result in data loss, since other machines have a full copy of the same dataset.

However, a master-follower architecture cannot guarantee full availability in the case of hardware failure. In particular, if the master instance fails, the application will be unable to write any data until the master is restored, or one of the followers is promoted to become the new master. The process of promoting a new master can be automated using built-in database features or third-party tools, but there will still be some downtime during which the application cannot write data.

Replication without a master

Cassandra solves this problem by simply removing the master instance from the picture. In Cassandra, when a piece of data is written, the write is sent to all of the nodes that should hold a copy of that data; no single node is authoritative. This neatly solves the availability problem: with no master instance, there is no single point of failure. If a node becomes unavailable, the data intended for it is still written to the other nodes that should store it; the application need not halt writing data.

In fact, Cassandra is even more robust when a node is unavailable to receive a write. Through a process called **hinted handoff**, other nodes in the cluster will store information about the write request, and then replay that request to the missing node when it becomes available again.

Returning to our model of Cassandra replication from the previous section, we can now expand it to account for replication. In particular, each virtual node is in fact stored on multiple physical nodes. Let's pick up our four-node cluster, taking a look at how data would be distributed with a replication factor of three:

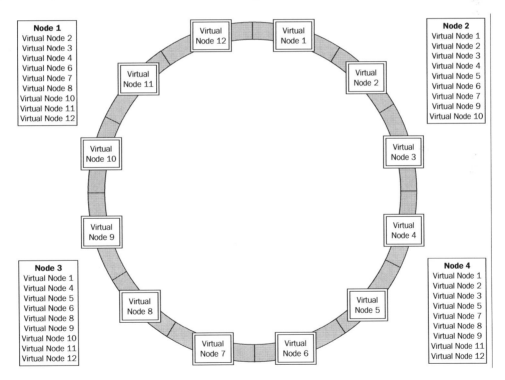

Each physical machine now claims responsibility for nine virtual nodes; each virtual node lives on three machines. If one machine fails, Cassandra can continue serving read and write requests, since copies of each of the failed machine's virtual nodes also live on two other machines. Even better, since virtual nodes are assigned randomly, none of the three healthy machines have to bear the entire burden of handling the failed machine's requests:

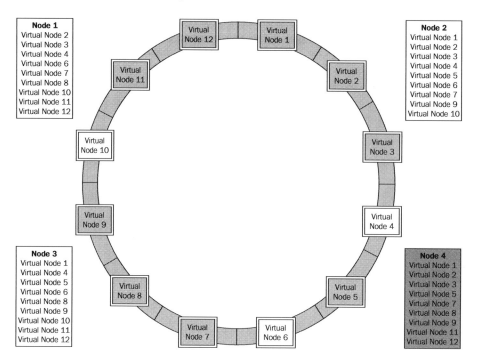

If the failed machine is permanently out of commission, it can be replaced by a new physical node; Cassandra will take care of populating the replacement machine with the data in the virtual nodes covered by the failed machine.

 Further readings on Cassandra's data distribution and replication can be found in the DataStax Cassandra documentation at `http://www.datastax.com/documentation/cassandra/2.1/cassandra/architecture/architectureDataDistributeAbout_c.html`.

Consistency

Masterless replication confers huge advantages in maintaining full availability in the face of hardware failure, but it also creates a thorny, if subtle, problem: how can we be sure that the data we're reading is the most recent version of that data?

Let's first consider the master-follower architecture we discussed above. When our application issues a write request to the master instance, that request only returns a successful response once the data has been successfully written to the master instance. A successful response does not, however, guarantee that the data has been replicated to all of the followers: any master-follower system will involve some delay before write are replicated to the followers.

Still, in this system, there is a simple way to guarantee that, when we read a piece of data, we're reading the most up-to-date version of it: simply read from the master. Of course, if we only ever read from the master, the followers don't do much for us other than providing a hot standby in case the master fails. But if we identify specific situations in which it's critical to have a guarantee that we're reading the most up-to-date data, we can read data from the master in only those situations, and make use of the followers in other scenarios.

Immediate and eventual consistency

In the master-follower scenario, we can say that reads from the master provide **immediate consistency**. Immediate consistency is simply a guarantee that, as soon as a piece of data is written successfully, it will be available to clients who are reading that data. There is no delay between writing data and being able to read it.

On the other hand, reads from the followers provide **eventual consistency**. Eventual consistency means that, once the data is successfully written, it will, at some point in the future, become available for reads. There is no guarantee of how soon it will become available; the only guarantee is that the time will eventually come. Of course, in a healthy production deployment, the delay between write and read availability is typically in the order of milliseconds, but we have no guarantee that it won't be much longer.

Consistency in Cassandra

As application developers using Cassandra, we also have a choice between immediate and eventual consistency. However, we don't implement this decision by selecting a particular machine to read from: when interacting with Cassandra, the particular node we're communicating with is irrelevant because requests are always forwarded to the node or nodes that should handle them.

Instead, Cassandra has a built-in notion of **tunable consistency**, which allows us to make a decision, for each query, about what consistency characteristics we want. Specifically, we tune consistency by telling Cassandra what constitutes a successful request.

The anatomy of a successful request

From the application's standpoint, when we issue a query to Cassandra, we simply send some CQL over the wire, and at some later point receive a response. Hopefully, the response will return successfully; if not, we will receive an error.

One requisite for a successful response is that our query is well-formed: if we misspell the name of a column, for instance, we will certainly see an error. But this condition for success is of little concern to us, since an invalid query will always generate an error, regardless of external factors.

More interesting from the perspective of consistency is the other condition of a successful request: that Cassandra is internally able to fulfill it. For a write request, this means durably storing the data being written; for a read request, it means retrieving the data from storage and returning it.

The crux of the question is this: How many nodes must be involved in the fulfillment of a request for it to be considered successful? For instance, if we are writing data to a row, we might consider the write successful as soon as the data is written to a single replica of that row; or we might require that the data be written to all of the replicas it belongs on before we consider it a success.

When reading data, the same logic applies: we may want Cassandra to just read one copy of the data and return it to us; or we may want it to read all copies of the data, compare their timestamps, and return to us the most recent version.

Tuning consistency

Cassandra's tunable consistency allows us to specify, for each query, how many replicas must be involved for the request to be considered a success. Consistency is not directly specified as part of the CQL query; instead, it is provided as an additional parameter given to the driver when performing a query. Different language drivers handle this differently, but any CQL driver should provide an interface for specifying the consistency for a given query.

Cassandra defines eleven different consistency levels, but many of them are for fairly specialized scenarios. We will explore the three most important and useful consistency levels: ONE, ALL, and QUORUM.

Eventual consistency with ONE

For the remainder of our discussion of consistency, let's assume we're using a replication factor of three. That means, regardless of the total number of nodes we have in our cluster, any given piece of data will live on three different nodes.

Let's say we update alice's row in the users table, changing her email address using the ONE consistency level. This means the write should be considered a success as soon as one copy of the data is written durably. When the request completes, we know that alice's new email address has definitely been written to one node, but we don't know whether the change has propagated to the other two replicas.

If, immediately after writing alice's row, we then attempt to read it back, also using ONE consistency, we're telling Cassandra that it can read any one of the three copies of the user row in the cluster, and return that copy to us. For the sake of illustration, suppose that at the exact moment we try to read the row, the replicas are in the following states:

- Replica 1 is the replica that acknowledged the write request. By definition, it has the up-to-date email address.
- Replica 2 has also written the updated email address durably.
- Replica 3 has not yet received the updated email address.

By reading with ONE consistency, we're saying that we're fine receiving any of the three replicas' copies of the user row. If the copy we get back happens to come from Replica 1 or Replica 2, we'll get back alice's record with her up-to-date email address. If it comes from Replica 3, however, we'll get an outdated record with her old address.

This is a temporary state of affairs: eventually, Replica 3 will also get the update, and at that point we will get the latest copy of the data regardless of which replica serves the request. But we don't know when that will be: when we read and write data using consistency level ONE, we are in a scenario of eventual consistency.

On the other hand, by using consistency level ONE, we're very resilient to failure. Even if two of the three machines that hold copies of alice's user record are unavailable, Cassandra can still successfully fulfill requests at this consistency level using the one healthy machine. As long as we're willing to give up immediate consistency, we can sleep very soundly knowing that our application will continue working even in the case of a multi-node hardware failure.

Eventual consistency is also beneficial for performance. When reading or writing data at ONE consistency, Cassandra doesn't only send the request to one replica; instead, it sends the request to all the interested replicas, and then fulfills the request as soon as the first one responds successfully. So the time it takes to service a request at ONE consistency is bounded by the fastest node to respond.

Immediate consistency with ALL

In many scenarios, however, eventual consistency is unacceptable. alice may be using the MyStatus web interface to update her email address. After submitting the form with her updated information, alice's browser is redirected to a page where she can view her profile. At this point, the web application will need to read the profile data from Cassandra; if it happens to read from an out-of-data replica, it will appear to alice that her email address was not actually updated.

At the other end of the consistency spectrum from ONE lies the ALL consistency level. At this level, all interested replicas must respond successfully in order for the request to be considered a success. If we update alice's email address with the ALL consistency, our UPDATE query won't complete successfully until the new email address has been written durably to all three replicas. As soon as the request completes, we know that all copies are up-to-date.

Similarly, if we read the row back with the ALL consistency, Cassandra will read all three replicas' copies of her user record, and compare their timestamps. If the data is different on different replicas, Cassandra will return the copy with the most recent timestamp, ensuring that we are getting the most recently written copy of the row.

Using the ALL consistency for both reading and writing, however, is an overkill in virtually all cases. If we write the data at the ALL consistency, then we can subsequently read it with ONE consistency because we know all of the replicas have the most recent copy of the data. Conversely, if we are reading with the ALL consistency, we will have immediate consistency even if the profile was updated with ONE consistency. At least one of the replicas has the most recent version of the data, and since we're reading all the copies, we're guaranteed to get the latest version back in one of them.

The downside of the ALL consistency is that, when we use it, we have no tolerance for failure. If any one of the machines that should store alice's user profile is unavailable, the request at the ALL consistency will fail. Queries at the ALL consistency also have unpleasant performance characteristics. Since Cassandra must wait for all interested replicas to respond, the overall performance of the request is bounded by the response time of the slowest replica.

Fortunately, there is a middle ground that allows us to have immediate consistency without completely sacrificing performance and resilience to failure.

Fault-tolerant immediate consistency with QUORUM

If we consider a request successful once it's been fulfilled by a single replica, we're very resilient to node failure, but we have to be willing to accept eventual consistency. If we require that every replica respond to a request before that request can complete, we have immediate consistency at the cost of making every replica a single point of failure. But we can achieve both immediate consistency and failure tolerance by defining a successful request as one that's fulfilled by *just enough* replicas.

This is the QUORUM consistency level, and *just enough* is precisely defined as the strict majority of replicas. For our replication factor of three, QUORUM requires that two nodes fulfill the request. Let's explore how the QUORUM consistency gives us the best of both worlds.

Returning to our email update scenario, let's send the query to change alice's email using the QUORUM consistency. Immediately after the UPDATE query completes, the worst-case scenario is as follows:

- Replica 1 is one of the replicas that acknowledged the write request. By definition, it has the up-to-date email address.

- Replica 2 is the other replica that acknowledged the write request, so it also has the up-to-date email address.

- Replica 3 has not yet received the updated email address.

By the contract of the QUORUM consistency level, we are assured that at least two of the three replicas have persisted alice's new email address. If we then read alice's user profile, also at QUORUM consistency, Cassandra will wait to receive at least two copies of the profile from two replicas. In the above scenario, at least one of the copies must come from Replica 1 or Replica 2; we're guaranteed that Cassandra will see the most recent version of the user profile before returning it to us.

This observation generalizes to quorums at any replication factor: as long as data is written to a strict majority of nodes and read from a strict majority of nodes, we know that there is *some* replica that was involved in both the write and the subsequent read. So, we're guaranteed immediate consistency.

> Note that we are only guaranteed immediate consistency if data is written and read at the QUORUM consistency. This is in contrast to the ALL consistency, which guarantees immediate consistency if it's used for either writing or reading.

The QUORUM consistency is also resilient to failure: if one of the three nodes holding a copy of alice's user profile is unavailable, Cassandra can still successfully fulfill requests using the other two nodes. QUORUM is not an entirely free lunch, however; operations at the ONE consistency level can handle a failure of any two nodes, whereas the QUORUM consistency can only handle a single node failure.

QUORUM performance is also a middle ground between ONE and ALL: QUORUM operations are bounded by the median response time of the interested replicas. This is a big advantage over the ALL consistency, if one replica happens to be under especially heavy load. But is not as speedy as ONE, which will be as responsive as the fastest node.

Comparing consistency levels

Having explored the three most useful consistency levels Cassandra provides, let's compare their important characteristics:

Level	Consistency behavior	Replicas required	Failure tolerance	Performance
ONE	Eventual	1	Tolerates failure of all but one replica	Fastest node
ALL	Immediate	All	None	Slowest node
QUORUM	Immediate, if used for both writes and reads	Strict majority	Tolerates failure of a minority of replicas	Median performing node

Cassandra supports eight other consistency levels; their uses are more esoteric and many deal specifically with deployments where a cluster is spread across multiple data centers. For a full reference, see `http://www.datastax.com/documentation/ cassandra/2.1/cassandra/dml/dml_config_consistency_c.html`.

Choosing the right consistency level

We've explored how the process of updating `alice`'s email address would work at the consistency levels `ONE`, `ALL`, and `QUORUM`. But what consistency levels should the MyStatus application actually use?

We recognized earlier that, in the case where `alice` updates her own user profile and then views the results, eventual consistency is undesirable: if she sees an outdated copy of her profile, it may appear that her email address update did not work at all. So, we will want to ensure that, from `alice`'s standpoint, changes to her profile are immediately consistent.

We've seen that, of the two levels that ensure immediate consistency, `QUORUM` has considerably more appeal from the perspective of fault-tolerance and performance. We also know that, in order to get immediate consistency from `QUORUM`, we'll need to both write and read at the `QUORUM` level.

So, when `alice` updates her email address, we'll write that column to Cassandra at `QUORUM`. When `alice` views her own profile, we'll also read her profile data from the users table at the `QUORUM` consistency. By always using `QUORUM` when `alice` is interacting with her own profile, we guarantee that she'll see the effects of all the changes she has made.

When other users are viewing `alice`'s profile, however, we have no strong need for immediate consistency. Here, `bob` will not know or care if the data he's seeing on `alice`'s profile is a few milliseconds out-of-date. So, eventual consistency is perfectly acceptable for the case of viewing another user's profile.

So, when displaying one user's profile to a different user, we'll use the more permissive `ONE` consistency. These requests will be more resilient to node failures, and will also perform better and impose less load on the cluster, since they won't require coordinating reads across multiple replicas.

For a deeper dive into consistency in Cassandra, see the DataStax Cassandra documentation at `http://www.datastax.com/ documentation/cassandra/2.1/cassandra/dml/ dmlAboutDataConsistency.html`.

The CAP theorem

Through tunable consistency, Cassandra allows us to choose where in the tradeoff between consistency and failure-tolerance we are most comfortable. But the tradeoff is there nonetheless; this is an unavoidable fact not just about Cassandra, but about all distributed systems. The tradeoff between consistency and availability is formalized by the **CAP Theorem**.

The CAP Theorem concerns the following three properties of a distributed system, from which it derives its name:

- Consistency in the CAP theorem refers to immediate consistency — the effects of a successful write operation must be immediately available to any client who wishes to read the data

- Availability refers to the ability of the system to give every request a response about whether it succeeded or failed

- Partition tolerance is resilience to a particular kind of hardware failure — the arbitrary loss of data in inter-node communication

The CAP theorem states simply that it is impossible for a distributed system to provide all three of the above properties, but that it is possible to provide any two of them. The CAP theorem is best treated as a framework from which to think about the properties of distributed systems, rather than a rigorous description of real-world databases. That said, we can briefly explore where Cassandra lies within the CAP framework.

When using ONE consistency, Cassandra is clearly an AP system; we don't get immediate consistency, but the cluster can tolerate a complete partition between nodes. If the node that receives a request is able to communicate with one of the interested replicas, it will give a successful response; otherwise, it will fail. Note that returning a failed response does not violate availability as defined in CAP; the only requirement is that the request terminate with information about whether it succeeded.

When using the QUORUM or ALL consistency, Cassandra becomes a CA system. Immediate consistency is guaranteed but only when all nodes in the cluster are able to communicate with one another. In the event of a partition of the cluster, any request that requires responses from nodes on both sides of the partition will fail.

The slides for the presentation that introduced the CAP conjecture (later the CAP theorem) can be found at http://www.cs.berkeley. edu/~brewer/cs262b-2004/PODC-keynote.pdf.

Handling conflicting data

As we explored above, Cassandra's masterless replication can lead to situations in which multiple versions of the same record exist on different nodes. Since there is no master node containing the canonical copy of a record, Cassandra must use other means to determine which version of the data is correct.

This situation comes into play when reading data at any consistency level other than ONE. When our application requests a row from Cassandra, we will receive a response with that row's data; each column will contain one value. However, if we're reading at a consistency level such as QUORUM or ALL, Cassandra internally will fetch the copies of the data from multiple nodes; it's possible that the different copies will contain conflicting data. It's up to Cassandra to figure out exactly what to return to us.

The problem is most acute when different clients are writing the same piece of data concurrently. Let's return to a scenario we explored in *Chapter 7, Expanding Your Data Model*: two employees of HappyCorp, Heather and Charles, are simultaneously attempting to update the location field in the user record of HappyCorp's shared account. Let's suppose that we are writing data at consistency level ONE. This concurrent operation could be carried out via the following sequence of events:

1. Heather updates the location to New York. The update is acknowledged by Replica 1.

2. Charles updates the location to Palo Alto. The update is acknowledged by Replica 2.

Just after Heather and Charles's concurrent updates, the Replica 1 copy of the HappyCorp user record will contain New York in its location field, and the Replica 2 copy will contain Palo Alto. Now, before the updates have a chance to propagate to any nodes except the ones that respectively acknowledged them, let's read the data back at the ALL consistency.

When Cassandra receives the read request, it will fetch HappyCorp's user record from Replicas 1, 2, and 3. Each replica will contain a different version of the record: Replica 1's copy has New York in the location field, Replica 2's has Palo Alto, and Replica 3's does not contain anything in that field. So what location will Cassandra actually return to us?

Last-write-wins conflict resolution

Under the hood, every time a piece of data is written to Cassandra, a timestamp is attached. Then, when Cassandra has to deal with conflicting data as in the scenario mentioned earlier, it simply chooses the data with the most recent timestamp. Even though the sequence of writes we discussed earlier was concurrent from the perspective of the distributed database, it's vanishingly unlikely that they were received by Cassandra at the exact same microsecond. So, one of them will have the more recent timestamp, and that's the one that Cassandra will return when we read the data back with the ALL consistency.

It's important to emphasize that each column has its own timestamp value; if we issue a query that only updates the location field in HappyCorp's user record, the location field will carry the timestamp of that operation, but the other fields in HappyCorp's record will still carry the timestamps of whenever they were last updated.

The ability to discretely update individual columns in a row, and have the timestamp of the last write associated with the column rather than the whole row, is why Cassandra is able to use the relatively simplistic last-write-wins strategy for conflict resolution. If, in the previous scenario, Heather had updated HappyCorp's location, but Charles had updated the email address, Cassandra would simply take the most recent version of each column, synthesizing an up-to-date view of the row that does not yet exist on any individual replica.

For an illuminating comparison of Cassandra's conflict resolution with strategies used by other distributed databases, see the blog post *Why Cassandra doesn't need vector clocks* at http://www.datastax.com/dev/blog/why-cassandra-doesnt-need-vector-clocks.

Introspecting write timestamps

As it turns out, CQL gives us a mechanism for finding out when a particular column was last written. This is done using the WRITETIME function. Let's say we're interested in the write timestamps for the email and location fields in HappyCorp's user record:

```
SELECT "email", WRITETIME("email"),
   "location", WRITETIME("location")
FROM "users"
WHERE "username" = 'happycorp';
```

We'll see that Cassandra gives us the write times of each column formatted as microseconds since the UNIX epoch:

```
email                  | writetime(email) | location     | writetime(location)
-----------------------+------------------+--------------+----------------------
 media@happycorp.com   | 1409607324125000 | New York, NY |    1409607607205000
```

Note that the timestamps are different: we initially wrote the email address when we created HappyCorp's record and later updated the row to have a location.

> For more information on timestamp introspection in Cassandra, see the DataStax CQL documentation at `http://www.datastax.com/documentation/cql/3.1/cql_using/use_writetime.html`.

Overriding write timestamps

Typically, write timestamps are assigned to data automatically by Cassandra, reflecting the time at which the cluster initially receives the request. However, CQL gives us a mechanism to override the default timestamp and instead choose our own.

To demonstrate, let's say we want to add a `location` value to `bob`'s user record at a specific timestamp: October 18, 2014 at 12:15:14.908323 EDT. In order to do this, we can use the USING TIMESTAMP clause in our UPDATE query:

```
UPDATE "users"
USING TIMESTAMP 1414953847665102
SET "location" = 'Austin, TX'
WHERE "username" = 'bob';
```

Reading `bob`'s user record back, we can see that it now has the location we specified:

```
 username | location
----------+------------
      bob | Austin, TX

(1 rows)
```

Now let's issue another update to bob's location, using a timestamp one microsecond before the one we provided previously:

```
UPDATE "users"
USING TIMESTAMP 1414953847665101
SET "location" = 'Tulsa, OK'
WHERE "username" = 'bob';
```

Now let's check the state of bob's row:

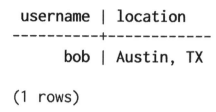

```
username | location
----------+------------
     bob | Austin, TX

(1 rows)
```

Notably, the location is still Austin. This leads to an important observation about write timestamps: as well as giving Cassandra information about resolving conflicts at read time, they're also used to determine at write time whether a given value is more up-to-date than the value Cassandra currently has stored. This is true even in a single-replica, single-node configuration such as the one we have set up for development.

Since the timestamp of our second update was earlier than the timestamp of our first update, Cassandra recognizes that the original value is actually a more *recent* version of the data, and thus discards the second update altogether.

Distributed deletion

When we want to delete data from storage, we might assume that Cassandra simply removes the data from disk and forgets that it ever existed. In a non-distributed environment, this approach to deletion would be entirely sufficient, but deletion is a bit more complex in a distributed database like Cassandra. To find out why, let's return to, and modify, our previous scenario with Heather and Charles making concurrent modifications to HappyCorp's user record.

In our modified scenario, Heather will still be updating the location column to contain New York, but Charles will be attempting to delete the contents of that column altogether. As with the original scenario, they'll be making their respective changes at roughly the same time, and our application will issue the UPDATE and DELETE queries at the ONE consistency level. Consider the following sequence of events:

1. Heather issues a request to update the location value to New York; this is acknowledged by Replica 1.

2. Charles issues a request to delete the contents of the location column; this is acknowledged by Replica 2.

Given that Charles's request to delete the location happened just after Heather's request to update it, the correct state of the location column is to contain no data. But with a naïve approach to deletion, that won't be the outcome. Let's consider what will happen if we read the HappyCorp user record at the ALL consistency just after Charles and Heather's requests complete, but before the requests have propagated to replicas other than the ones that respectively acknowledge them.

When the coordinator reads all copies of the HappyCorp user record, it will see one version with a location field containing New York, one version with no data in the location column, and one version with the old location prior to either of their updates. Observing that the version with the most recent timestamp contains New York, it will determine that this is the up-to-date value of the location field. This is a violation of immediate consistency, however, since the last change to the location column was the deletion, the ALL-consistency read should reflect that fact.

To deal with scenarios like this, Cassandra does not completely forget that a value ever existed when it is deleted. Instead, it stores a **tombstone** in place of the deleted value; that tombstone is simply a marker that data was deleted from that column. Tombstones, like normal column values, carry a timestamp indicating when the deletion happened.

Armed with a tombstone, Cassandra can correctly serve our request to read HappyCorp's user record just after the concurrent update and deletion. The node coordinating the read will now see a version of the row with New York in the location field, a version of the row with the old location, and a version of the row with the tombstone in the location column. Since the tombstone will have the most recent write timestamp of the three, the coordinator will correctly determine that there should be no data returned for the location field.

For more on deletion and tombstones, see the DataStax Cassandra documentation at `http://www.datastax.com/documentation/cassandra/2.1/cassandra/dml/dml_about_deletes_c.html`.

Stumbling on tombstones

We've seen that tombstones are critical to ensure that Cassandra can correctly identify that a piece of data has been deleted in a distributed environment, but tombstones also have a downside. Since tombstones are stored in place of the deleted values, they continue to occupy space in the range of clustering columns in a given partition. In some situations, this can lead to unexpected performance degradation and even errors.

To use a somewhat artificial illustration, let's say that `alice` is now a long-time MyStatus user, and has created tens of thousands of status updates. Let's also assume that `alice` decides one day that she wants to delete 1,000 recent status updates she's created. Once she's done with that process, we have stored a thousand tombstones, each of which is stored at the ID of the deleted status update in the table's data structure.

The next time someone wants to read `alice`'s user timeline, we're going to run into a slight problem. If we ask for 20 most recent status updates in her timeline, Cassandra will dutifully scan the 20 most recent IDs in `alice`'s partition. But it will find that these are all tombstones! Luckily, Cassandra is smart enough to keep looking; it won't simply return an empty result set. But it will have to scan a thousand tombstones before it finds any non-deleted data. This scan carries a substantial performance overhead when compared to the same operation with no tombstones getting in the way.

For that reason, it's best to avoid situations in which your application will need to scan over clustering column ranges containing lots of tombstones.

Tombstones do not live forever. Instead, they are automatically cleared by Cassandra's compaction process after a configured amount of time has elapsed. By default, that duration is ten days. This means that a node may rejoin the cluster after up to ten days of downtime without a risk of deleted records springing back to life. From our perspective as application developers, the relatively long life of tombstones means that we are best off thinking of them as permanent, and strenuously avoiding situations where many tombstone records must be scanned in order to service a query.

Expiring columns with TTL

We first encountered the CQL DELETE command, which allows us to remove the value of a column or an entire row, in *Chapter 5, Establishing Relationships*. Another way to remove data from Cassandra is to give it a **time-to-live** (**TTL**). This is equivalent to deleting data in the future: when we write data with a TTL, it is marked as deleted after the amount of time we specify has elapsed.

CQL allows us to attach a TTL to any INSERT or UPDATE statement. Let's experiment with setting an expiring location value:

```
UPDATE "users"
USING TTL 30
SET "location" = 'Vancouver'
WHERE "username" = 'bob';
```

TTL values are given in seconds, so what we've done here is tell Cassandra that bob's location should be Vancouver for the next thirty seconds, after which it should be deleted. CQL also allows us to introspect the TTL attached to any data column using the TTL function:

```
SELECT "username", "location", TTL("location")
FROM "users"
WHERE "username" = 'bob';
```

We'll see that the TTL column in the result tells us how many seconds remain until the location field expires:

```
 username | location  | ttl(location)
----------+-----------+---------------
      bob | Vancouver |            17

(1 rows)
```

If we now wait the full thirty seconds and then check bob's user record again, we'll see that the location value has disappeared.

Expiring columns can be useful in situations where we don't want data to stick around forever. For instance, we may want to generate a password reset link using a random token that is emailed to a user. We might want that token to only work for 24 hours, to reduce the risk that it's used by someone else besides the user requesting the password reset. Expiring columns give us a seamless way to accomplish this.

Summary

In this chapter, we confronted and explored the major issues that stem from Cassandra's masterless, distributed, and replicated architecture. Interacting with Cassandra often feels indistinguishable from working with a single-node data store but, when working with any distributed database, we need to think about the tradeoff between consistency and availability. In some situations, we might be willing to read slightly out-of-date data for the sake of performance and failure tolerance; in others, we will tolerate a higher probability of a request failing in order to ensure that the data we're reading is fully up-to-date.

You learned that the partition key for a row determines not only its physical location in storage, but also which nodes within the cluster store copies of the row. This further motivates the practice of designing our table schemas such that most queries are looking for data grouped under a single partition key.

You discovered how, in a distributed database, deletion is not as simple as removing a value from storage and forgetting it ever existed. We saw that maintaining tombstones for deleted data allows Cassandra to guarantee that deletions are always reflected when reading a row with strong consistency. We also saw that tombstones can lead to performance degradation in scenarios where we're trying to scan clustering column ranges that contain many deleted rows.

In the first *Appendix A, Peeking Under the Hood*, we will peek under Cassandra's hood, exploring how the advanced functionality of CQL is in fact a rich abstraction on top of a much simpler underlying data structure. Peeking under the hood is often a great deal of fun; and if you're interested in exploring the nuts and bolts of Cassandra data structures, I encourage you to read on. But the material in the *Appendix A, Peeking Under the Hood* is purely supplementary: you are fully prepared to be an effective user of Cassandra without it.

A
Peeking Under the Hood

Over the previous ten chapters, we've thoroughly explored Cassandra's capabilities from the perspective of application developers. All of our interaction with Cassandra has been through CQL, and we've explored a robust set of features available to us via the CQL interface. We've found that one of the big appeals of Cassandra is the rich set of data structures available to us for domain modeling; structures such as compound primary keys, collection columns, and secondary indexes are part of what sets Cassandra apart from other distributed databases.

As it turns out, CQL is an abstraction on top of a much less sophisticated data structure that underlies all the data stored in Cassandra. Commonly referred to as the **Thrift interface**, named after the protocol used to interact with Cassandra at this level, this layer represents all data using an ad hoc key-value structure called a **column family**. As developers, we will never need to interact with Cassandra at this level of abstraction, but it's illuminating to explore how the familiar CQL data structures are represented at the column family level.

In some contexts, you might hear the terms table and column family used interchangeably. In fact, CQL provides CREATE COLUMNFAMILY as an alias for the familiar CREATE TABLE command. This is a legacy of the gradual divergence of the CQL table structure from the underlying column family structure. In earlier versions of CQL, tables were mapped more or less transparently onto the underlying column family structure.

We will use the terminology preferred by the Cassandra developers, using table to talk about the CQL data structure and column family for the lower level structure available via the Thrift interface.

When we have completed exploring the Thrift interface, you'll have discovered:

- How to directly access Cassandra's low-level column family structures
- Column family representation of simple and compound primary keys
- The structure of collection columns at the column family level

Using cassandra-cli

To explore the Thrift API, we'll use a new tool called **cassandra-cli**. Like cqlsh, cassandra-cli is a command-line interface to Cassandra, but it does not provide a CQL interface. Instead, cassandra-cli uses a small, purpose-built query language that allows us to interact directly with column families. Some cassandra-cli commands resemble their CQL equivalents, but it's merely a resemblance, not a relationship.

In recent versions of Cassandra, cassandra-cli is deprecated and will be removed from Cassandra 3.0. This should serve to underscore the fact that CQL is considered the way to interact with Cassandra: the Thrift interface is merely a curiosity, not a viable tool to interact with Cassandra in our applications.

Your installation of Cassandra should have the cassandra-cli executable in the same directory as cqlsh. Once we start it up, we should see something like this:

```
$ cassandra-cli
Connected to: "Test Cluster" on 127.0.0.1/9160
Welcome to Cassandra CLI version 2.0.7

The CLI is deprecated and will be removed in Cassandra 3.0.  Consider migrating to cqlsh.
CQL is fully backwards compatible with Thrift data; see http://www.datastax.com/dev/blog/thrift-to-cql3

Type 'help;' or '?' for help.
Type 'quit;' or 'exit;' to quit.

[default@unknown] 
```

Just as in cqlsh, we'll need to start by specifying the keyspace we want to interact with:

```
USE my_status;
```

Now that we've got cassandra-cli up and running, we can start exploring how our CQL3 tables are modeled at the column family level.

The structure of a simple primary key table

To start with, let's have a look at the `users` table. To do this, we'll start with the `LIST` command that prints all the data in a given column family:

```
LIST users;
```

This will print out a long list of information, grouped by `RowKey`. For brevity, the first couple of `RowKey` groups appear as follows:

```
RowKey: bob
=> (name=, value=, timestamp=1401412616531000)
=> (name=encrypted_password, value=10920941a69549d33aaee6116ed1f47e19b8e713, timestamp=1401412616531000)
=> (name=version, value=ec0c1fb1321f11e48eeb5f98e903bf02, timestamp=1409607341355000)
-------------------
RowKey: happycorp
=> (name=, value=, timestamp=1409607324125000)
=> (name=email, value=6d6564696406861707079636f72702e636f6d, timestamp=1409607324125000)
=> (name=encrypted_password, value=368200fa910c16cc644f3512e63b541c85fa2a3c, timestamp=1409607324125000)
=> (name=location, value=4e657720596f726b2c204e59, timestamp=1409607607205000)
=> (name=version, value=8a810110322011e48eeb5f98e903bf02, timestamp=1409607607205000)
```

Although we've never seen it structured like this before, the data here should look pretty familiar. The `RowKey` headers correspond to the username column in our CQL3 table structure. Within each `RowKey` is a collection of tuples, each tuple containing a name, a value, and a timestamp. We will call these tuple cells, in keeping with the terminology used in the cassandra-cli interface itself.

 You might encounter the word `column` being used for the name-value-timestamp tuples we are exploring here. Not only does that terminology invite confusion with the concept of a column in CQL3, but it's also a singularly misleading way to describe the data structure in question. We'll stick with "cell", which is both unambiguous and more descriptive. Please excuse us for referring to collections of cells as column families—there is no better alternative.

Exploring cells

Looking at the `name` attribute, we see things like `email`, `encrypted_password`, `version`, and `location`. Clearly, the `name` attribute of the cells corresponds to the names of columns in our CQL schema—although the relationship is more complex than it might appear, as we'll explore in the next section.

The `value` field in the cells is a bit of a mystery; given that the name contains column names, we might expect that the value would contain column values. However, what we see in cassandra-cli are just some inscrutable hexadecimal blobs.

As it turns out, under the hood Cassandra represents all data as hexadecimal byte arrays; the type system is part of CQL's abstraction layer. The cassandra-cli utility does give us a way to retrieve human-readable values of individual columns, using the `AS` keyword to explicitly specify the type. Let's try to read the value of the cell with the `name` value as `email`, from HappyCorp's user record:

```
GET users['happycorp']['email'] AS ascii;
```

The `GET` command allows us to access a single cell by first specifying the `RowKey` value, then the cell name. We can also omit the cell name to return all the cells at a given `RowKey`.

In this case, thanks to our use of `AS`, we can see a human-readable value for HappyCorp's email address:

```
=> (name=email, value=media@happycorp.com, timestamp=1409607324125000)
Elapsed time: 3.34 msec(s).
```

Thankfully, the CLI will remember our preference for reading emails in ASCII, and will accordingly print out all cells named `email` from the `users` column family for the remainder of the session.

A model of column families: RowKey and cells

At this point, we've satisfied ourselves that each cell in a column family corresponds to a column name and value at the CQL level. The mapping between CQL3 tables and lower-level column families, so far, seems pretty straightforward. A table row's primary key value is stored in the `RowKey` field, and each cell contains a name-value pair that represents the value of the named column in the row.

We can make the relationship between the table representation and the column family representation more accessible with a visualization of each:

The **users** table

username	email	encrypted_password	location
bob		0x10920941...	
happycorp	media@happycorp.com	0x368200fa...	New York, NY

The **users** column family

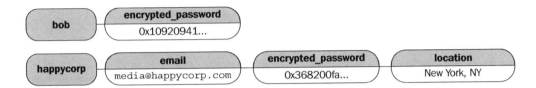

We thus visualize the column family as a collection of RowKey values, with each RowKey linked to a collection of cells containing a name and a value. Note that not every CQL3 column is represented by a cell in every RowKey field; only when a given column has a value does it have a corresponding cell in the column family. This is consistent with our mental model of CQL3 tables, developed in the *Developing a mental model for Cassandra* section of *Chapter 2, The First Table*.

As it turns out, the similarities between column families and CQL3 tables take us this far and no further.

Compound primary keys in column families

Now that we've established the relatively familiar-looking column family structure of users—a table with a simple primary key—let's move on to a table with a compound primary key. To start, let's take a look at home_status_updates, a fairly straightforward table. Recall that this table has a partition key timeline_username; a clustering column status_update_id; and two data columns, body and status_update_username column.

We'll use the LIST command to take a look at the contents of the column family and, beforehand, we'll use the ASSUME command to set the value output format to utf8. This has a similar effect as the AS modifier we used earlier, but it applies to all cells in a column family, rather than only cells with a specific name:

```
ASSUME home_status_updates VALIDATOR AS utf8;
LIST home_status_updates;
```

The output of the LIST command takes the same general shape as that for the users column family, but the way the information is arranged might come as a surprise:

```
RowKey: alice
=> (name=16e2f240-2afa-11e4-8069-5f98e903bf02:, value=, timestamp=1408823428694000)
=> (name=16e2f240-2afa-11e4-8069-5f98e903bf02:body, value=dave update 4, timestamp=1408823428694000)
=> (name=16e2f240-2afa-11e4-8069-5f98e903bf02:status_update_username, value=dave, timestamp=1408823428694000)
=> (name=cacc7de0-2af9-11e4-8069-5f98e903bf02:, value=, timestamp=1408821757668000)
=> (name=cacc7de0-2af9-11e4-8069-5f98e903bf02:body, value=carol update 4, timestamp=1408821757668000)
=> (name=cacc7de0-2af9-11e4-8069-5f98e903bf02:status_update_username, value=carol, timestamp=1408821757668000)
```

The output contains only a single RowKey, that is alice; it contains six cells with rather inscrutable names. Recall that in the *Fully denormalizing the home timeline* section of *Chapter 6, Denormalizing Data for Maximum Performance*, we populated the home_status_updates table with two status updates, both in alice's partition of the table. So how does the information in the column family line up with the information in the CQL table we're used to?

To answer this question, look closely at the names of the cells. Take the second cell, which has the name 16e2f240-2afa-11e4-8069-5f98e903bf02:body. This cell name appears to contain two pieces of information: a clustering column value and the name of a data column. The value of the cell is, straightforwardly, the value of the body column in the row with the status_update_id value 16e2f240-2afa-11e4-8069-5f98e903bf02.

The reason we only see one RowKey value is because all the rows in the home_status_updates table have the partition key alice. What cassandra-cli calls a RowKey is, in fact, exactly what we have been calling a partition key the entire time.

A complete mapping

With this insight, we can now generate a complete mapping between the components of a column family and the CQL-level table structure components they represent:

Column Family Component	CQL3 Table Component
RowKey	Partition Key
Cell name	The values of clustering column(s), if any, followed by the name of the data column
Cell value	The value in the row/column identified by the partition key, clustering column(s), and the column name

To reinforce this idea, let's apply our visualization from the previous section to the more complex structure of `home_status_updates`:

The **home_status_updates** table

username	status_update_id	body	status_update_username
alice	16e2f240...	dave update 4	dave
alice	cacc7de0...	carol update 4	carol

The **home_status_updates** column family

alice	16e2f240...:body	16e2f240...:status_update_username	cacc7de0...:body	cacc7de0...:status_update_username
	dave update 4	dave	carol update 4	carol

The main insight here is that every value in every row in `alice`'s partition is stored in a one-dimensional data structure in the column family. At the column family level, nothing in the storage structure distinguishes the values in `carol`'s status update row from the values in `dave`'s status update row. The cells containing those values simply have different prefixes in their cell names; only at the CQL level are these represented as distinct rows.

The wide row data structure

Another important observation is that, at the column family level, there is no preset list of allowed cell names. The cell names we observe in `alice`'s partition of the `home_status_updates` column family are derived partially from data calculated at runtime—namely, the UUID values of various status update ID columns. At the CQL level, tables must have predefined column names but no equivalent restriction exists at the column family level.

Looking closely at the cells in `alice`'s RowKey, we observe that they are in order: first, descending by the timestamp encoded in the UUID component of the cell name, and second, alphabetically ascending by the column name component. You learned in *Chapter 3, Organizing Related Data* that clustering columns give CQL rows a natural ordering, and now we know how the cells grouped under a given RowKey at the column family level are always ordered by the cell name.

The data structure that we observe at `alice`'s RowKey in `home_status_updates` is commonly called a **wide row**. Not to be confused with the rows we commonly interact with in CQL tables, wide rows are simply ordered collections of cells; each wide row is associated with a RowKey, which at the CQL level we call a partition key.

The empty cell

We've so far ignored one curious entity in the column family data structure—each `RowKey` seems to have a cell whose value is empty, and whose name consists purely of a clustering column value and a delimiter.

Recall that, in a CQL row, there is no requirement that any data column contain data. A row consisting purely of values for the primary key columns is entirely valid. CQL makes this possible by storing an extra cell any time a row is created or updated that acts as a placeholder indicating that there is a row in the clustering column encoded in the cell's name. Thus, the existence of the row is guaranteed even if there is no data column to be stored as a cell in that row.

Collection columns in column families

We've seen the surprising way in which the key structure of a CQL table maps to the underlying column family representation, but so far the values stored in CQL have mapped completely transparently into cell values at the column family level. While this is true for scalar data types, something much more interesting happens where collections are concerned.

Set columns in column families

We'll start by looking at sets, which are the simplest of the three collection column types. Let's take a look at `alice`'s row in the `user_status_updates` column family:

```
GET user_status_updates['alice'];
```

Recall that, by only using the bracket operator once in the GET statement, we'll retrieve all cells in the `alice` RowKey.

There are quite a few cells in `alice`'s wide row, but we're particularly interested in those having to do with the `starred_by_users` column, which is a set:

```
=> (name=76e7a4d0-e796-11e3-90ce-5f98e903bf02:starred_by_users:626f62, value=, timestamp=1411314062536000)
=> (name=76e7a4d0-e796-11e3-90ce-5f98e903bf02:starred_by_users:6361726f6c, value=, timestamp=1411314212388000)
```

Note that the cell names here look like the cell names in `home_status_updates`, but with yet another component: a short hexadecimal string appended to the end. Unfortunately, there's no straightforward way to get cassandra-cli to print those hexadecimal strings in a human-readable form, so you'll have to take my word for it that `0x626f62` is the blob encoding of the string `bob`, and `0x6361726f6c` encodes the string `carol`.

With this cleared up, we can observe that the underlying representation of a CQL set is a collection of cells, with each cell encoding one of the items in the set. The item is actually encoded in the cell's *name* by appending it to the usual combination of clustering column value and row name that comprises the name of a scalar cell.

Let's have a look at a set using the visual comparison we have been developing in the course of this chapter. For clarity, I've omitted everything unrelated to the set:

The **user_status_updates** table

username	status_update_id	starred_by_users
alice	76e7a4d0...	{dave, carol}

The **user_status_updates** column family

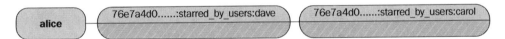

This turns out to be a very effective way to represent a set at the column family level. When a new item is added to a set, the CQL abstraction simply adds a cell containing the item appended to the appropriate clustering column/column name combination. Removing items from the set is as simple as deleting the cell that encodes that item. In both cases, Cassandra can do the right thing without having to read the existing contents of the set.

Map columns in column families

Although maps might seem like the most complex of the three collection column types, their column family level representation is actually simpler than that of lists, so we'll save lists for last. To get a sense of what map columns look like at the column family level, let's take a look at the social_identities map in `alice`'s user record:

```
GET users['alice'];
```

Again, we'll focus solely on the interesting cells in the result:

```
=> (name=encrypted_password, value=8914977ed729792e403da53024c6069a9158b8c4, timestamp=1409605884231000)
=> (name=social_identities:74776974746572, value=2725634, timestamp=1411314494120000)
```

The structure of the cell names looks identical to the cell names used by set columns. However, unlike set columns, the cells encoding map columns have values in them. In a map, the map keys are stored in cell names, and the map values are stored in cell values. We can thus visualize a map like this:

The **users** table

username	social_identities
alice	{'twitter':2725634, 'yo':25}

The **users** column family

The relationship between the underlying representations of map columns and set columns should come as no great surprise; many programming languages implement their set data type using an underlying map whose values are not meaningful.

Our reasoning about the process of insertion and deletion in sets also carries over to maps. Since, for any given map key that needs to be inserted or deleted, the CQL abstraction can calculate the corresponding cell name, map key-value pairs can be directly inserted and deleted without Cassandra needing to read the full contents of the column.

List columns in column families

The last collection type we'll explore at the column family level is lists. Lists are the most complex collection types in a couple of ways. As you learned in *Chapter 8, Collections, Tuples, and User-defined Types*, lists have more available mutation operations than maps or sets. It also turns out that lists have the most complex underlying representation.

Let's take a look at the `shared_by` list in the `user_status_updates` table. Before we dive into the underlying structure, let's quickly open a cqlsh console and add a new value to the `shared_by` list of `alice`. When we last left this list, it only had a single username in it, which isn't really useful for the purposes of our exploration:

```
UPDATE "user_status_updates"
SET "shared_by" = "shared_by" + ['bob'],
   "shared_by" = ['dave'] + "shared_by"
WHERE "username" = 'alice'
AND "id" = 76e7a4d0-e796-11e3-90ce-5f98e903bf02;
```

Note that, by setting the `shared_by` column twice in our UPDATE statement, we can both prepend and append values to the list in a single query. Having done this, let's take a look at `alice`'s partition in the `user_status_updates` column family, focusing on the cells related to the `shared_by` collection:

```
=> (name=76e7a4d0-e796-11e3-90ce-5f98e903bf02:shared_by:4116d73041a611e4b4ad5f98e903bf02,
value=726f62657274, timestamp=1411314365135000)
=> (name=76e7a4d0-e796-11e3-90ce-5f98e903bf02:shared_by:e358efe0678711e4ac53f51fa8d64e2e,
value=626f62, timestamp=1415479406045000)
```

At first glance, the structure looks quite similar to that of set and map columns: the cell name consists of a clustering column value, a data column name, and finally some other list-related component. However, what is the list-related component?

It turns out that this value is a UUID; the hex string printed in cassandra-cli is in fact the usual UUID representation, just without the dashes. If we decode the UUID `e358efe0678711e4ac53f51fa8d64e2e`, the last one in the list, we'll find that it encodes the timestamp at which we appended `bob` to the list column.

Appending and prepending values to lists

Our observation allows us to deduce that Cassandra appends items to list columns by creating a new cell whose name includes the current timestamp UUID. Since the value of timestamp UUIDs is monotonically increasing, this will ensure that the appended value is the last in the list, since it has the highest (newest) UUID. This way, Cassandra can construct the cell name for a newly appended list item without consulting the current contents of the list.

Now, let's take a look at the first cell in the list, the one whose value contains `dave`. Recall that we prepended `dave` onto the list, so we know that the cell name doesn't contain the current timestamp because that would put Dave at the end of the list, not the beginning.

Decoding the UUID in Dave's cell's name, `77d9443f854711d9ac53f51fa8d64e2e`, yields a timestamp several years old — February 23, 2005. Where does this timestamp come from? As it turns out, the following calculation yields it:

- Calculate how much time has elapsed since midnight UTC on January 1, 2010

- Calculate the timestamp corresponding to that duration before January 1, 2010

So while the UUIDs used to append to lists are monotonically increasing, the UUIDs used to prepend are monotonically decreasing, since we'll be subtracting an ever-greater duration from the "epoch" time of January 1, 2010. Since the first version of Cassandra to support lists was released in early 2013, we can assume that that particular epoch marker was chosen to ensure that all generated prepend timestamps would be smaller than all generated append timestamps.

Other list operations

We've seen that the structure of maps and sets allows Cassandra to perform all supported mutation operations on those collection types without having to read the currently stored values. Unfortunately, the same is not true for lists. While appending and prepending to lists can be done discretely, other operations require scanning the existing contents of the list first.

If we take a moment to ponder the underlying representation of list columns, the reason is clear. If, for instance, we ask to remove the value `dave` from the list, there's no way for Cassandra to know which cell(s) contain that value without reading it. Even in the simpler case of deleting an element at a given position, Cassandra must look at the cell names to know which cell is at the position requested. So, it's worth keeping in mind that list mutations — except for appending and prepending — are naturally less performant than mutations on other collection types.

Summary

Having spent the bulk of this book exploring Cassandra using CQL, we've now peeked under the hood to see how the robust data structures offered by CQL are in fact abstractions upon a much more rudimentary column family data structure. If nothing else, this is an opportunity to appreciate modern technology; as recently as Cassandra 1.1, the CQL data structures we take for granted were unheard of. Before Cassandra 0.7, CQL did not exist at all; the Thrift interface we explored in this chapter was state-of-the-art.

Our deep dive into Cassandra's internal data structures also helps us understand the underlying reasons behind facts that in previous chapters we'd taken as first principles. You learned in *Chapter 3, Organizing Related Data* that partition keys group related data, but now we know that in fact all the data under a partition key, or `RowKey`, coexists in a single wide row. Similarly, we already knew that clustering columns determine row ordering, but now we know that the sorting is actually happening to cells in a wide row, and the use of the clustering column as a prefix for wide row cell names generates the familiar behavior at the CQL level.

In the final appendix, we will do a brief survey of Cassandra's authentication and authorization functionality, and discuss the various tools Cassandra gives you for protecting your deployments against security vulnerabilities.

B
Authentication and Authorization

In our interactions with Cassandra in this book, we haven't concerned ourselves with authentication or authorization; whenever we connect to our local Cassandra instance, we're not required to provide any credentials, and there have been no restrictions on what kind of operations we've been able to perform. This is the default configuration for a Cassandra cluster and works well in many scenarios, in particular where network access to the machines running Cassandra is tightly controlled.

In some scenarios, however, it's useful to be able to control access to Cassandra at the database level itself—for instance, when a cluster is shared between multiple tenants or where a large organization needs to restrict access to sensitive data to certain departments or individuals.

For these scenarios, Cassandra does offer a full suite of authentication and authorization functionality; accounts and permissions are configured using CQL. In order to enable authentication and authorization in our development cluster, we will need to make a couple of small changes to the cluster's configuration file, but we'll otherwise be operating in the familiar territory of cqlsh.

By the end of this appendix, you'll be familiar with:

- How to configure your cluster to restrict access to authorized users
- How to create a user
- How to change a user's password
- How to grant privileges to a user
- How to see the privileges granted to a user
- How to revoke privileges from a user
- Other steps you may want to take to secure your Cassandra cluster

Enabling authentication and authorization

By default, Cassandra does not require user authentication when clients connect to the cluster, and it also does not place any restriction on the ability of clients to perform operations on the database. To change this, we will need to make a couple of minor modifications to our Cassandra instance's configuration file. Since modifications to the configuration file are typically a concern of deployment engineers, we haven't interacted with it in this book, so you may be wondering where to find it. Where it's located depends on your platform; the table below assumes you installed Cassandra using the instructions for your platform in the *Installing Cassandra* section of *Chapter 1, Getting Up and Running with Cassandra*. The following table gives you the location of the `cassandra.yaml` file on the respective platforms.

Platform	Configuration file location
Mac OS X	`/usr/local/etc/cassandra/cassandra.yaml`
Ubuntu	`/etc/cassandra/cassandra.yaml`
Windows	`C:\Program Files\DataStax Community\apache-cassandra\conf\cassandra.yaml`

You will need to make two changes to the `cassandra.yaml` file. First, find the line that begins with `authenticator:` and change it to:

```
authenticator: PasswordAuthenticator
```

This change tells Cassandra to require a username and password when clients connect to the cluster. It does not, however, restrict access based on which user is logged in; to do that, we'll need to enable authorization. Find the line beginning with `authorizer:` and change it to:

```
authorizer: CassandraAuthorizer
```

Now our cluster will restrict the access of the logged in user based on the permissions that user has been granted. You will need to restart your Cassandra instance for the settings to take effect.

Authentication, authorization, and fault tolerance

Using the `PasswordAuthenticator` and `CassandraAuthorizer` strategies for authentication and authorization respectively, user credentials and granted permissions are stored in Cassandra itself. This means that, if the authentication data becomes unavailable, no clients will be able to access the cluster. For that reason, you will always want to set the replication factor for your `system_auth` keyspace to the total number of nodes in your cluster. Since our development cluster consists of only a single node, we don't need to make any changes; in production, however, you will almost certainly have many nodes in your cluster, and you'll want to make sure credentials and permissions are stored locally on every one of them.

For further information on configuring Cassandra authentication, including best practices when configuring authentication in a production cluster, see the DataStax Cassandra documentation at `http://www.datastax.com/documentation/cassandra/2.1/cassandra/security/security_config_native_authenticate_t.html`.

For more information on configuring authorization, see `http://www.datastax.com/documentation/cassandra/2.1/cassandra/security/secure_config_native_authorize_t.html`.

Authentication with cqlsh

Now that we've enabled authentication on our development cluster, we will need to reconnect our cqlsh session with a username and password. By default, Cassandra has a superuser account whose username and password are both `cassandra`, so we can use that:

```
$ cqlsh -u cassandra -p cassandra
```

In a production cluster, of course, we would not want to have a superuser account with such an easily guessable password; however, for our purposes, this will work fine.

Authentication in your application

The details differ from language to language, but the CQL driver for your platform should provide a mechanism for authenticated connections. Consult your driver's documentation for more information.

Setting up a user

By default, Cassandra has one user account configured; its username is `cassandra` and it is a superuser, meaning there are no restrictions on what it can do to the database. When you connect as a superuser, it's essentially the same as if Cassandra did not have authorization enabled at all.

In order to do something useful with our access restrictions, we'll want to set up another user account that is not a superuser. Let's say one of the departments in the burgeoning MyStatus corporation is a data analytics team. This team needs access to read data from our Cassandra cluster, but has no need to add, change, or remove data. We'll set up a user account for this team that gives them only the access they need.

First, we'll use the `CREATE USER` command to add a user account for the analytics team:

```
CREATE USER 'data_analytics'
WITH PASSWORD 'strongpassword'
NOSUPERUSER;
```

Note, of course, that in a real deployment we would want to choose an actual strong password, for which the string `strongpassword` does not suffice.

The `NOSUPERUSER` option tells Cassandra that the new account is not a superuser, meaning it can't do anything it hasn't explicitly been granted permission to do. This is the default for new users, but we make it explicit here for clarity; replacing it with `SUPERUSER` would make the new user a superuser.

> For more information on the `CREATE USER` command, see the DataStax CQL documentation: `http://www.datastax.com/documentation/cql/3.1/cql/cql_reference/create_user_r.html`.

Changing a user's password

If we wish to change the password of an existing account, we can use the ALTER USER command:

```
ALTER USER 'data_analytics'
WITH PASSWORD 'verystrongpassword';
```

We can also append a SUPERUSER or NOSUPERUSER option to the end of this query to grant or revoke superuser access.

Viewing user accounts

We can see existing user accounts, along with their superuser status, by accessing the users table in the system_auth keyspace:

```
SELECT * FROM "system_auth"."users";
```

Note that the use of the period between the keyspace name and the table name simply instructs CQL that we'd like to look into the system_auth keyspace, regardless of which keyspace, if any, has been activated by an USE statement.

As we can see, user accounts are stored quite transparently in the system_auth. users table:

```
     name           | super
----------------+-------
        cassandra   | True
   data_analytics   | False

(2 rows)
```

You may be wondering where and how passwords are stored. These live in a separate table, system_auth.credentials; passwords are not stored in plain text, but rather as bcrypt hashes.

Controlling access

We've now created a user account for our data analytics team, but so far that user can't actually do anything in our database. We'd like to give the data analytics team read access to all the data in our application's keyspace, but no ability to modify data or schema structures. To do this, we'll use the GRANT command:

```
GRANT SELECT PERMISSION
ON KEYSPACE "my_status"
TO 'data_analytics';
```

Now the data_analytics user can read the data from any table in the my_status keyspace; however, it can't make any modifications to anything. The SELECT permission we used above is one of six that Cassandra makes available:

Permission	Description	CQL commands allowed
SELECT	Read data	SELECT
MODIFY	Add, update, and remove data in existing tables	INSERT UPDATE DELETE TRUNCATE
CREATE	Create keyspaces and tables	CREATE TABLE CREATE KEYSPACE
ALTER	Modify the structure of existing keyspaces and tables	ALTER TABLE CREATE INDEX DROP INDEX ALTER KEYSPACE
DROP	Drop existing keyspaces and tables	DROP TABLE DROP KEYSPACE
AUTHORIZE	Grant or revoke permissions to other users	GRANT REVOKE

There is also an ALL permission that is simply shorthand for granting all six of the permissions listed above.

As well as granting a permission across an entire keyspace, we can also grant permissions to operate on a specific table. For instance, we might want to give the data analytics team full read/write access to the `status_update_views` table that we created in *Chapter 9, Aggregating Time-Series Data*, since the data in that table is primarily an analytics concern. To do that, we'll use GRANT to issue a permission on a specific table:

```
GRANT MODIFY PERMISSION
ON TABLE "my_status"."status_update_views"
TO 'data_analytics';
```

Note that we use ON TABLE instead of ON KEYSPACE here, and that we provide the fully-qualified name of the table we want the permission to apply to. Of course, if we've previously issued a USE "my_status"; command in this session, we can refer to the table name without providing the keyspace name explicitly.

> The GRANT statement can be used to provide all manner of access, from a very broad mandate to a very narrow set of permissions. For full details of the GRANT syntax, see the DataStax CQL reference: http://www.datastax.com/documentation/cql/3.1/cql/cql_reference/grant_r.html.

Viewing permissions

Much like user accounts, permissions are stored in a human-readable way in a Cassandra table, in this case the permissions table. We can see the effects of the permissions that we granted by reading from this table:

```
SELECT * FROM "system_auth"."permissions";
```

We'll see a row for each of the permissions we've granted:

```
 username        | resource                             | permissions
-----------------+--------------------------------------+-------------
 data_analytics  |                      data/my_status  | {'SELECT'}
 data_analytics  | data/my_status/status_update_views   | {'MODIFY'}

(2 rows)
```

Note that the `resource` column contains a path to the keyspace or column family that the user has permissions on, and that the permissions column is in fact a set column. Note also that the `cassandra` user does not appear in this list; as a superuser, the `cassandra` user automatically has permission to do anything it wants, and does not need explicit authorization of the kind stored in the permissions table.

Revoking access

Let us say that, having slept on it, we've reconsidered the decision to give the data analytics team read/write access to the `status_update_views` table. After all, the data in that table is generated automatically by the application, and there's no need for the analytics team to be able to modify it. Happily, we can revoke that permission just as easily as we granted it, using the REVOKE command:

```
REVOKE MODIFY PERMISSION
ON "my_status"."status_update_views"
FROM 'data_analytics';
```

The syntax of the REVOKE command is identical to that of the GRANT command, except that the TO keyword is replaced by FROM. We can check the `system_auth`. `permissions` table to satisfy ourselves that the permissions are how we'd like:

```
 username        | resource        | permissions
-----------------+-----------------+-------------
 data_analytics  | data/my_status  | {'SELECT'}

(1 rows)
```

We can rest easy knowing that the only remaining access granted to the analytics team is keyspace-wide read-only permissions.

 For more on REVOKE, see the DataStax CQL documentation: http://www.datastax.com/documentation/cql/3.1/cql/cql_reference/revoke_r.html.

Authorization in action

Now that we know how to create user accounts and grant and revoke permissions to them, let's see how a non-superuser account behaves in practice. To do this, let's open up a new cqlsh session logged in with our data analytics team's account:

```
$ cqlsh -u data_analytics -p verystrongpassword -k my_status
```

The `-k my_status` option simply tells cqlsh that we want to interact with the `my_status` keyspace, saving us the effort of issuing a `USE` statement.

Now let's see what we can do. First, we expect to be able to read data with no problem; let's have a look at the `user_status_updates` table:

```
SELECT * FROM user_status_updates;
```

As expected, we have permission to read the contents of that table:

```
 username | id                                   | body
----------+--------------------------------------+--------------------------
      bob | 97719c50-e797-11e3-90ce-5f98e903bf02 | Eating a tasty sandwich.
      bob | e93a7410-7344-11e4-a0e6-5f98e903bf02 |           Bob Update 1
      bob | e93bac90-7344-11e4-a0e6-5f98e903bf02 |           Bob Update 2
      bob | e9454980-7344-11e4-a0e6-5f98e903bf02 |           Bob Update 3
    alice | 76e7a4d0-e796-11e3-90ce-5f98e903bf02 |     Learning Cassandra!
    alice | e9367c70-7344-11e4-a0e6-5f98e903bf02 |         Alice Update 1
    alice | e93b3760-7344-11e4-a0e6-5f98e903bf02 |         Alice Update 2
    alice | e9443810-7344-11e4-a0e6-5f98e903bf02 |         Alice Update 3

(8 rows)
```

Now let's try making a change to some data. Though our analytics team certainly would have no malicious intent, perhaps at some point the analytics cat may sit on a keyboard, producing the following statement:

```
DELETE FROM "users"
WHERE "username" = 'alice';
```

That's quite an alarming query, but happily our authorization setup has saved us:

```
code=2100 [Unauthorized] message="User data_analytics has no MODIFY
permission on <table my_status.users> or any of its parents"
```

Recall that the `MODIFY` permission is needed to make any changes to existing data, including insertion and deletion. Since the `data_analytics` account only has the `SELECT` permission, our accidental attempt to delete `alice`'s account is rejected. `alice`'s data is safe.

Authorization as a hedge against mistakes

We generally think of authentication and authorization as a mechanism to prevent intentional access to our data by nefarious actors. However, authorization can also be a powerful insurance policy against unintentional mistakes by well-intentioned people. In the preceding example, the data analytics team did not intend to do any harm, but without authorization in place, that pesky cat would have unwittingly caused data loss.

While the odds of a feline posterior producing a perfectly-formed CQL query are quite long, mistakes do happen. Using authorization to give each user the minimum level of access they strictly need, we can reduce the chance of a mistake turning into an emergency.

Of course, authentication and authorization are also an important tool to secure your data from those seeking unauthorized access. As it turns out, these are only part of the entire security picture; we also need to make sure our data is secure on disk and in transit.

Security beyond authentication and authorization

The security afforded by Cassandra-level authentication and authorization only applies to clients connecting directly to your Cassandra cluster. Anyone who has physical access to the machines running Cassandra can access the data stored on disk; the same goes for anyone with SSH access to machines in the Cassandra cluster. Cassandra itself does not offer encryption for on-disk data, but DataStax Enterprise, a commercial distribution of Cassandra, does offer encryption of at-risk data. For more information, consult `http://www.datastax.com/documentation/datastax_enterprise/4.5/datastax_enterprise/sec/secTDE.html`.

Data security can also be compromised in transit; anyone who can intercept traffic between your application and your Cassandra cluster can potentially gain unauthorized access to your data. Cassandra offers client-to-node SSL encryption that protects your data in transit between your application and your cluster. For information on setting up client-to-node encryption, see `http://www.datastax.com/documentation/cassandra/2.1/cassandra/security/secureSSLClientToNode_t.html`.

Finally, normal operation of Cassandra involves passing data between different nodes in the cluster; if attackers can intercept inter-node communication, they can gain access to your data. Cassandra has the ability to encrypt all node-to-node traffic; for information on configuring node-to-node encryption, see `http://www.datastax.com/documentation/cassandra/2.1/cassandra/security/secureSSLNodeToNode_t.html`.

Security protects against vulnerabilities

We've covered authentication and authorization in depth and also mentioned a few other security measures available in Cassandra or DataStax Enterprise. However, not every deployment will need every security measure in place. What security measures you need depends on the sensitivity of your data and the security characteristics of your deployment. As an application engineer, you likely leave the details of deployment security to your sysadmins, but this table can provide a quick reference to what sort of security measures you should think about:

Security measure	Vulnerability protected against	Alternative protections
Internal authentication & authorization	Direct access to Cassandra cluster via CQL binary protocol	Restrict access to the Cassandra cluster to secure the private network
On-disk data encryption	Shell access to machines hosting the Cassandra cluster	Restrict shell and physical access to Cassandra hosts
Client-to-node encryption	Network link between the application and the Cassandra cluster	Restrict access to private network traffic using encryption or physical isolation
Node-to-node encryption	Network link between Cassandra nodes	Restrict access to private network traffic using encryption or physical isolation

It's quite plausible that, with both your application and Cassandra deployment protected by a well-secured VPN, you may not need to concern yourself with any database-level security measures.

Summary

In this final appendix, we explored Cassandra's ability to restrict access to itself using internal authentication and authorization. We saw that Cassandra offers simple configuration of user accounts and permissions using a collection of CQL commands provided for that purpose, and also that this information is stored transparently in tables in the `system_auth` keyspace.

We noted that internal authorization can be useful for traditional security concerns, but also simply as a hedge against mistakes. By limiting access to that which is strictly needed, we can reduce our vulnerability to user errors that can unintentionally cause major data loss.

We also noted that internal authentication and authorization are not the full security picture for a Cassandra deployment. While the details are beyond the scope of a book whose audience is primarily application engineers, we did a brief survey of other security measures a Cassandra deployment might undertake, and we briefly discussed situations in which those measures might be called for.

Wrapping up

The first nine chapters of this book were devoted to learning CQL's major features and how to effectively use them. *Chapter 10, How Cassandra Distributes Data*, explored the special issues that application developers must keep in mind when using a distributed database such as Cassandra. In *Appendix A, Peeking Under the Hood*, we took a look at the low-level data structures that were once part of the daily lives of Cassandra developers but now serve as an internal data structure underlying the more powerful abstractions of CQL. And, in this appendix, you learned about the tools Cassandra gives you to keep your deployment secure.

This brings us to the end of the topics we set out to cover in this book. You should now be fully prepared to take advantage of Cassandra's powerful feature set, high scalability, and resilience to failure in your own applications. You know that Cassandra is better suited to some use cases than others; when a distributed database is called for, you're prepared to use the most powerful distributed database on the planet, efficiently and productively.

Index

follow relationships, storing
about 77
denormalization 78
queries, designing around 78

H

hinted handoff 175
home timeline
denormalizing 99
displaying 97-101
generating 90, 91
status update, creating 100, 101
horizontal scalability 8

I

immediate consistency
about 11
with ALL 180
inbound follows, follow
relationships 76, 77
inserts
about 113
checking 114, 115
existing data, overwriting 113, 114
INSERT statement 30
collections 138, 139
integers
bigint type 26
int type 26
varint type 26
int type 26

K

keyspace
about 19, 20
selecting 20
key-value pairs
storing, with maps 136

L

last-write-wins conflict resolution 186
lightweight transactions 116, 117, 122

list
elements, removing 135, 136
operations 204
using, for nonunique values 133
using, for ordered values 133
values, appending 203, 204
values, prepending 203, 204
writing 133
list column
defining 133
in column families 202, 203

M

map columns
in column families 201, 202
MapReduce 12
maps
discrete values, updating 137
used, for storing key-value pairs 136
values, removing from 138
writing 137
masterless replication 174, 176
multiple partitions
and read efficiency 93
paginating over 68-70

N

natural key 24
node 8
node-to-node encryption
URL 217
normalized approach
about 90
multiple partitions 93
ordering 92
pagination 92
read efficiency 93
timeline, generating 90, 91

O

ONE
eventual consistency 179

indexing 149
partial selection 149, 150
querying 149
URL 148
using 147
user table
about 24
properties 25
structuring 24
uuid type 27

V

values
appending, to lists 203, 204
prepending, to lists 203, 204
removing, from column 111
syntactic sugar, for deletion 112
varint type 26
virtual nodes 171, 172

W

WHERE keyword
clustering column, restricting by 60, 61
limitations 60
part of partition key, restricting by 61, 62
wide row
data structure 199, 200
Windows
Cassandra, installing on 18
write complexity
and data integrity 101-103
write timestamps
introspecting 186
overriding 187, 188

Thank you for buying
Learning Apache Cassandra

About Packt Publishing

Packt, pronounced 'packed', published its first book, *Mastering phpMyAdmin for Effective MySQL Management*, in April 2004, and subsequently continued to specialize in publishing highly focused books on specific technologies and solutions.

Our books and publications share the experiences of your fellow IT professionals in adapting and customizing today's systems, applications, and frameworks. Our solution-based books give you the knowledge and power to customize the software and technologies you're using to get the job done. Packt books are more specific and less general than the IT books you have seen in the past. Our unique business model allows us to bring you more focused information, giving you more of what you need to know, and less of what you don't.

Packt is a modern yet unique publishing company that focuses on producing quality, cutting-edge books for communities of developers, administrators, and newbies alike. For more information, please visit our website at www.packtpub.com.

About Packt Open Source

In 2010, Packt launched two new brands, Packt Open Source and Packt Enterprise, in order to continue its focus on specialization. This book is part of the Packt Open Source brand, home to books published on software built around open source licenses, and offering information to anybody from advanced developers to budding web designers. The Open Source brand also runs Packt's Open Source Royalty Scheme, by which Packt gives a royalty to each open source project about whose software a book is sold.

Writing for Packt

We welcome all inquiries from people who are interested in authoring. Book proposals should be sent to author@packtpub.com. If your book idea is still at an early stage and you would like to discuss it first before writing a formal book proposal, then please contact us; one of our commissioning editors will get in touch with you.

We're not just looking for published authors; if you have strong technical skills but no writing experience, our experienced editors can help you develop a writing career, or simply get some additional reward for your expertise.

Learning Cassandra for Administrators

ISBN: 978-1-78216-817-1   Paperback: 120 pages

Optimize high-scale data by tuning and troubleshooting using Cassandra

1.   Install and set up a multi datacenter Cassandra.

2.   Troubleshoot and tune Cassandra.

3.   Covers CAP tradeoffs, physical/hardware limitations, and helps you understand the magic.

4.   Tune your kernel, JVM, to maximize the performance.

Cassandra Design Patterns

ISBN: 978-1-78328-880-9   Paperback: 88 pages

Understand and apply Cassandra design and usage patterns, and solve real-world business or technical problems

1.   Learn how to identify real world use cases that Cassandra solves easily, in order to use it effectively.

2.   Identify and apply usage and design patterns to solve specific business and technical problems including technologies that work in tandem with Cassandra.

3.   A hands-on guide that will show you the strengths of the technology and help you apply Cassandra design patterns to data models.

Please check **www.PacktPub.com** for information on our titles

[PACKT] open source ✿

PUBLISHING   community experience distilled

Mastering Apache Cassandra

ISBN: 978-1-78216-268-1   Paperback: 340 pages

Get comfortable with the fastest NoSQL database, its architecture, key programming patterns, infrastructure management, and more!

1. Complete coverage of all aspects of Cassandra.

2. Discusses prominent patterns, pros and cons, and use cases.

3. Contains briefs on integration with other software.

Cassandra High Performance Cookbook

ISBN: 978-1-84951-512-2   Paperback: 310 pages

Over 150 recipes to design and optimize large-scale Apache Cassandra deployments

1. Get the best out of Cassandra using this efficient recipe bank.

2. Configure and tune Cassandra components to enhance performance.

3. Deploy Cassandra in various environments and monitor its performance.

4. Well illustrated, step-by-step recipes to make all tasks look easy!

Please check **www.PacktPub.com** for information on our titles